Oil and Gas in NZ

their effects on climate change

Jeremy King

Non-fiction: Oil and gas in NZ

ISBN 978-0-473-63742-2

Chapter		**Page**
1	Zealandia & Tectonic Plates	1
2	Geology of Fossil Fuels	5
3	Statistical Production of Crude Oil & Gas in NZ	10
4	Fossil Fuel Basins	15
5	Petroleum Fields	18
6	On-shore Production	22
7	Off-shore Production	25
8	Blow-outs	28
9	Hydraulic Fracturing	30
10	Cuttings	32
11	Royalties	33
12	Crude Oil & Gas Composition	34
13	Sulfur & Salt in Crude Oil	37
14	Refinery Production	40
15	Fuels used in NZ Transport	44
16	Gas Fuels	45
17	Oil Fuels	52
18	Coal Fuels	67
19	Hydrocarbon Energy & Vehicle Combustion	74

Chapter 1

Zealandia & Tectonic Plates

New Zealand is rich in untapped fossil fuel resources, and has a well developed crude oil, gas and coal infrastructure.

New Zealand, a land mass of just 268,830 km^2, lies on a mainly submerged 4,900,000 km^2 continental shelf, Zealandia, coined the world's lost, 8th largest continent.

The other 7 continents in order of size are Asia, Africa, North America, South America, Antarctica, Europe and Australia (and now mainly submerged Zealandia, the 8th largest).

Zealandia was once part of super-continent, Gondwanaland, which included the continents of Africa, South America, Antarctica and Australia.

Gondwanaland began splitting up 180 million years ago, and then Australia-Zealandia separated from each other some 100 million years later, forming the Tasman Sea between the two continents.

Zealandia was low-lying land with many swamps, and a unique ecosystem of flora and fauna evolved. As the climate was warmer and wetter than New Zealand of now, Zealandia was covered in sub-tropical rain forests, providing the fossil fuel of petroleum and coals that is available and used today.

Zealandia extends just past New Caledonia to the north and not too far off the shores of Australia.

Continents, like Australia and submerged Zealandia, are embedded in tectonic plates, massive slabs of rock that encase the earth as a crust, the lithosphere. The tectonic plates gradually move apart or collide together at the rate of between 2 to 15 centimetres per year, caused by convection currents of magma, hot molten rock below the lithosphere.

Over millions of years, the tectonic plate under Zealandia became thin due to plate separation movement, divergence, causing the continent of Zealandia to slowly submerge below the ocean.

Then, some 23 million years ago, subduction occurred, the colliding of the Australian Continental Tectonic Plate with the Pacific Oceanic Tectonic Plate, one plate sliding over the other, causing the land of New Zealand to be pinched upwards above sea-level.

As depicted on the previous page, all the North Island and the northern part of the South Island and west of the Southern Alps is embedded in the Australian Plate. The South Island, east of the Southern Alps, is embedded in the Pacific Plate.

The Southern Alps have been thrust up by the Australia Plate pushing under the Pacific Plate, subduction. The reverse is east of the North Island, the Pacific Plate is pushing under the Australian Plate.

New Zealand is the largest land area of submerged continental shelf Zealandia that is above sea-level, followed by the French territory of New Caledonia to the north, approximately 1,460 km from Cape Reinga, the northern most tip of New Zealand.

Zealandia slopes away from the land of New Zealand, and its far edge, the 'continental slope', is mostly about 3 kilometres below sea-level, then descends fairly sharply to the deep ocean floor, 5 kilometres plus in depth.

The maritime jurisdiction of Zealandia is split between New Zealand, Australia and New Caledonia.*

New Zealand has the largest share of the maritime jurisdiction in Zealandia, over that of Australia and New Caledonia.

New Zealand has sovereign rights to all resources on and under the seabed to that part of the continental shelf of Zealandia under its jurisdiction, that is, to all the crude oil and gas resources therein contained.

During a glacial ice age, the evaporation of water up into the atmosphere falls to become a build-up of snow and ice, causing sea-level to lower by as much as 120 metres, and as a consequence, when this occurs, the land of New Zealand stretches out into the sea, to be nearly twice the land area than of today. The last ice age ended about 25,000 years ago.

As the effect of an ice age is less severe in the Southern Hemisphere, and because New Zealand indigenous trees are hardy, most New Zealand forests are not lost during an ice age.

Thus, there has been a steady increase of matter depth of decayed vegetation and remains from animals over millions of years, including that from stretched-out New Zealand during a glacial period, now decomposed and formed into fossil fuel.

* Refer map on page 107 for the Treaty Line split of the submerged continental shelf, Zealandia, between New Zealand, Australia and New Caledonia.

Chapter 2

Geology of Fossil Fuels

Petroleum Formation

Petroleum, that is oil and gas, is usually formed below ocean seabeds, from the decomposition by single-celled anaerobic bacteria and archaea on oxygen-deprived marine organic matter, plus from vegetation and remains from animals flushed from land into the sea. Also formed from dense sub-tropical rain forests of Zealandia after submergence below sea-level.

The gradual accumulation of sediment above, sand, silt and clay of up to one thousand metres in depth, resulted in pressure on the decomposing organic matter beneath, and with temperature of up to 50°c from the decomposing organic matter transformed it into a waxy substance, kerogen.

With more heat, between 100 and 150°c from earth's geothermal heating, and from a greater pressure of sediment above, a few thousand metres in depth, the kerogen underwent the process of catagenesis, transforming the kerogen by pyrolysis (heating of organic matter in the absence of oxygen — a Greek-derived word, 'pyro', heat) into crude oil, composed of hydrocarbon molecules.

With further geothermal heating, between 150 and 220°c, gas was given off.

As petroleum is light, the crude oil and gas, by expulsion, migrated upwards through that sediment which had formed into porous (rock with voids) and permeable (allowing petroleum to pass through the rock) source rock, often sandstone, limestone or shale.

And where sediment above the source rock had formed into solid rock from compression, sedimentary impermeable rock, fossil fuels were trapped underneath to form underground reservoirs, sedimentary fossil fuel basins.

The composition of hydrocarbon molecules in fossil fuels determines the various consistency of oils and gases produced, reflecting in the defined end use. The composition depends upon the type of original marine and terrestrial organic matter decomposed, the amount of heat generated from decomposition and from geothermal activity, the pressure from the sedimentary impermeable rock above and the length of process over millions of years.

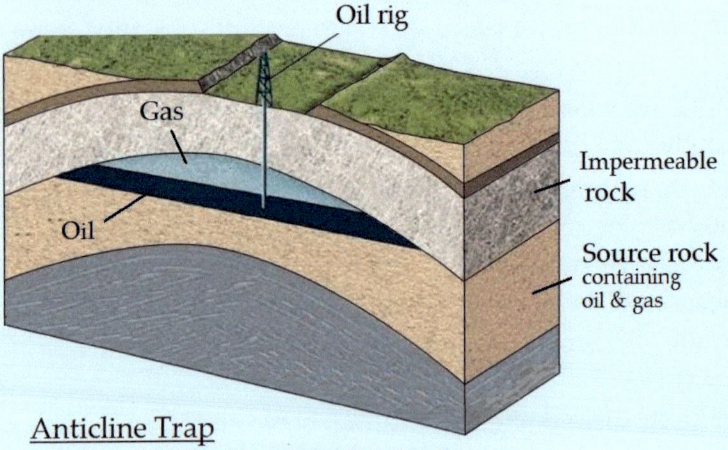

Anticline Trap
Oil & gas rise to the crest of the fold

Crude oil and gases were trapped in the reservoir basins by earth's shifting landmasses, in anticline traps, fault traps or stratigraphic traps, as depicted in the diagrams.

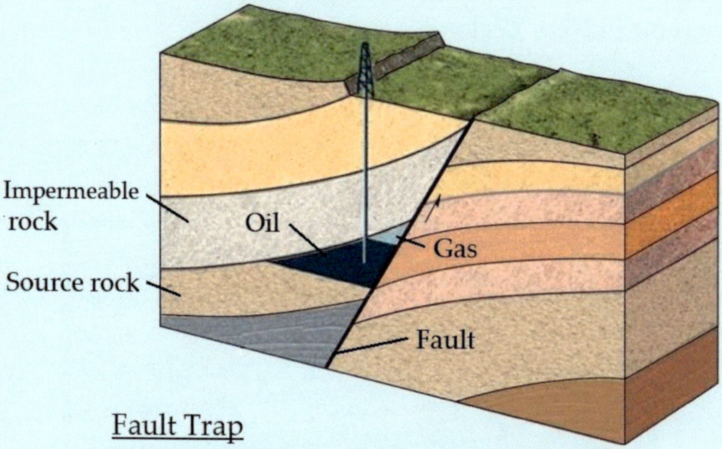

Fault Trap

Oil & gas collect in tilted strata next to fault

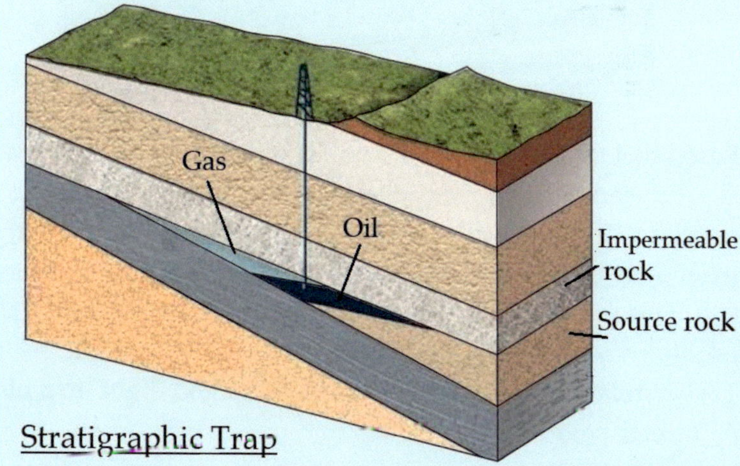

Stratigraphic Trap

Oil & gas collect where the source rock layer pinches it out

Occasionally, bubbling oil and gas reaches the ground surface through fissures, a 'seep', as discovered in 1865 by Edward Smith at Ngāmotu Beach on the New Plymouth foreshore of Taranaki, the first sign of a possible petroleum reservoir in New Zealand.

Later that year, a well was dug at Moturoa on the New Plymouth foreshore, and in 1866 the well struck gas at a depth of 7 metres and oil at 20 metres.

Post Card Moturoa Oilfields

Fossil fuel that is formed in New Zealand sedimentary basins is produced from vegetation and animal matter of over the previous 75 to 150 million years, more evolved matter than from most basins around the world, 280 to 360 million years ago. This more evolved vegetation in New Zealand has produced a higher quality, waxy 'sweet' and 'light' oil. 'Sweet' relates to less sulfur content in oil and 'light' to a high API (refer page 35).

The word, petroleum, is derived from Latin, 'petra' for rock and 'oleum' for oil.

Coal Formation

Coals are mostly derived from once swamp forests on coastal plains in New Zealand, or from dense sub-tropical rain forests of Zealandia before submergence below sea-level.

If the land became inundated with water, as then the water-logged soils were low in oxygen, the terrestrial matter of dead vegetation and remains from animals did not decay from aerobic bacteria, requiring oxygen for cellular respiration, and over millions of years the matter became many metres thick.

Anaerobic bacteria in the swamp forests, 'wetlands', not requiring oxygen, decomposed the terrestrial matter, emitting methane gas, 'swamp gas', transforming the matter into peat, partially decayed organic matter. *

Where sediment of sand, silt and clay accumulated over the peat, with the resultant increase in pressure and temperature, coalification occurred, the condensing of peat into coal by up to $^{1}/_{15}$ the volume of peat.

The increase of pressure and temperature in peat formed lignite, the lowest grade of coal, then to sub-bituminous, bituminous and lastly to the highest grade, anthracite (although this grade is not formed in New Zealand).

Refer page 67 for depth, to form different grades of coal.

* Where decomposition of matter is by aerobic bacteria, requiring oxygen for energy, less harmful carbon dioxide is emitted, although is a greenhouse gas.

Chapter 3

Statistical Production of Crude Oil & Gas in New Zealand

Crude oil and natural gas production is the extraction from petroleum reservoirs.

Tables in this chapter are compiled from the Ministry of Business, Innovation & Employment's (MBIE) 'Energy in New Zealand 2022' statistics (relating to the year 2021).

Renewable energy is geothermal, hydro, wind, bio-fuel and solar. Non-renewable energy is gas, oil and coal (although these fuels are renewable over millions of years).

Comparison of renewable and non-renewable energy in New Zealand:

	Renewable	Non- renewable
Energy Production *	57%	43%
Consumer Energy **	29%	71%

* **Energy Production** is indigenous energy produced from renewable and non-renewable sources. Significantly more renewable energy is produced in New Zealand than is non-renewable.

** **Consumer Energy** is after imports and exports and after 'transformation', the energy actually consumed by industry and by the public. Renewable energy consumed is less than 30% of total energy consumed.

'Transformation' is the generating of one form of energy from another. Electricity is a generated energy as depicted in the table below.

Comparison is in petajoules (Pj) of energy as the common denominator. 'What is petajoules?' — next page.

Electricity is generated from	**%**		**Pj's**
Geothermal	55%		195
Hydro-dams	25%		87
Wind	3%		10
Biogas	1%		3
Solar	<1%	84%	< 1
Natural gas	9%		30
Sub-bituminous coal	7%		25
Diesel	<1%	16%	< 1
Nuclear	0		0
Total		**100%**	**352**

Thus, 16% of energy in electric vehicles is from a non-renewable source, including in the 16% of which is high carbon content coal with greater pollutant emissions.

Australia's electricity transformation is from coal (75%) and gas (16%), a total of 91% from non-renewable fossil fuels.

One Pj is the equivalent energy produced by up to 30 million litres of regular petrol, or energy consumed per year by a city the size of Gisborne or Blenheim.

One petajoule is a quadrillion joules (one followed by 15 zeros – one million billion). One joule is the energy expended lifting an apple about one metre. One litre of petrol contains roughly 3.5 million joules of energy.

Energy Production

Natural gas and liquid petroleum gas (LPG) production in New Zealand during 2021 was nearly 4.5 billion m^3, producing 165 Pj of energy (as in the below table).

The volume of natural gas is measured at a temperature of 15°c and with a pressure of 100 kilopascals (kPa).

Crude oil production has been up to 20 million barrels per year, producing about 120 Pj of energy, although is now progressively declining, approaching 6 million barrels, 39 Pj of energy, during 2021.

The below tables show the energy sources from indigenous production in New Zealand during the year 2021, from both non-renewables and from renewables.

Non-Renewables

Energy Source	%	Pj
Natural gas	25%	157
Crude oil	6%	39
Bituminous coal	6%	38
Sub-bituminous coal	4%	27
Liquid petroleum gas	1%	8
Lignite coal	< 1%	5
Total	**43%**	**274**

Renewables

Energy Source	%	Pj
Geothermal	32%	204
Hydro-dams	14%	87
Bio-solid fuels	8%	50
Wind	2%	10
Bio-gas	< 1%	4
Solar	0	1
Total	**57%**	**356**

Bio-mass is organic plant (including wood) and animals, which produce both bio-solid fuel (the combustion of wood and herbaceous plants) and bio-gas (the gas given off by the decomposition of plant and animals — methane).

Total energy production during 2021, non-renewables and renewables, was 630 Pj.

During 2021, non-renewables — gas, crude oil and coal, were 43% of total indigenous energy produced, and this percentage appears to be gradually decreasing.

Consumer Energy

After imports and exports are taken into account and after 'transformation' of one energy source to another, consumer energy for the year, 2021:

Consumer Energy	% of Energy	Petajoules	
Diesel	27%	143	
Petrol	19%	103	
Natural gas (incld bio-gas)	12%	67	
Sub-bituminous coal	3%	14	
Aviation fuel	3%	13	
Liquid petroleum gas	1%	8	
Fuel oil (for ships)	1%	6	
Lignite coal	1%	5	
Bituminous coal	0	2	
Electricity transformed from fossil fuels	4%	23	**384**
Electricity transformed from renewables	23%	121	
Bio-solid fuels	5%	29	
Geothermal	1%	7	
Solar	0	< 1	
Wind	0	0	**158**
Total	**100%**		**542**

Non-renewable energy consumed, diesel, petrol, natural gas, coal, aviation fuel, LPG and fuel oil for the year 2021, was 71% of total energy consumed.

Chapter 4

Fossil Fuel Basins

New Zealand's 19 mainly off-shore sedimentary fossil fuel basins, covering an area of over 2 million km^2, have not yet been fully explored, and many may contain a number of potential commercially productive oil and gas reservoirs.

Pre-historic wetlands that produces fossil fuels

Exploration for crude oil and gas, 'wildcat' drilling, has occurred around New Zealand, although mainly in the Taranaki Basin. Drilling has also occurred in the Northeast Slope Basin, Reinga Basin, Pegasus Basin, East Coast Basin, Canterbury Basin and Great South Basin, depicted on the following page.

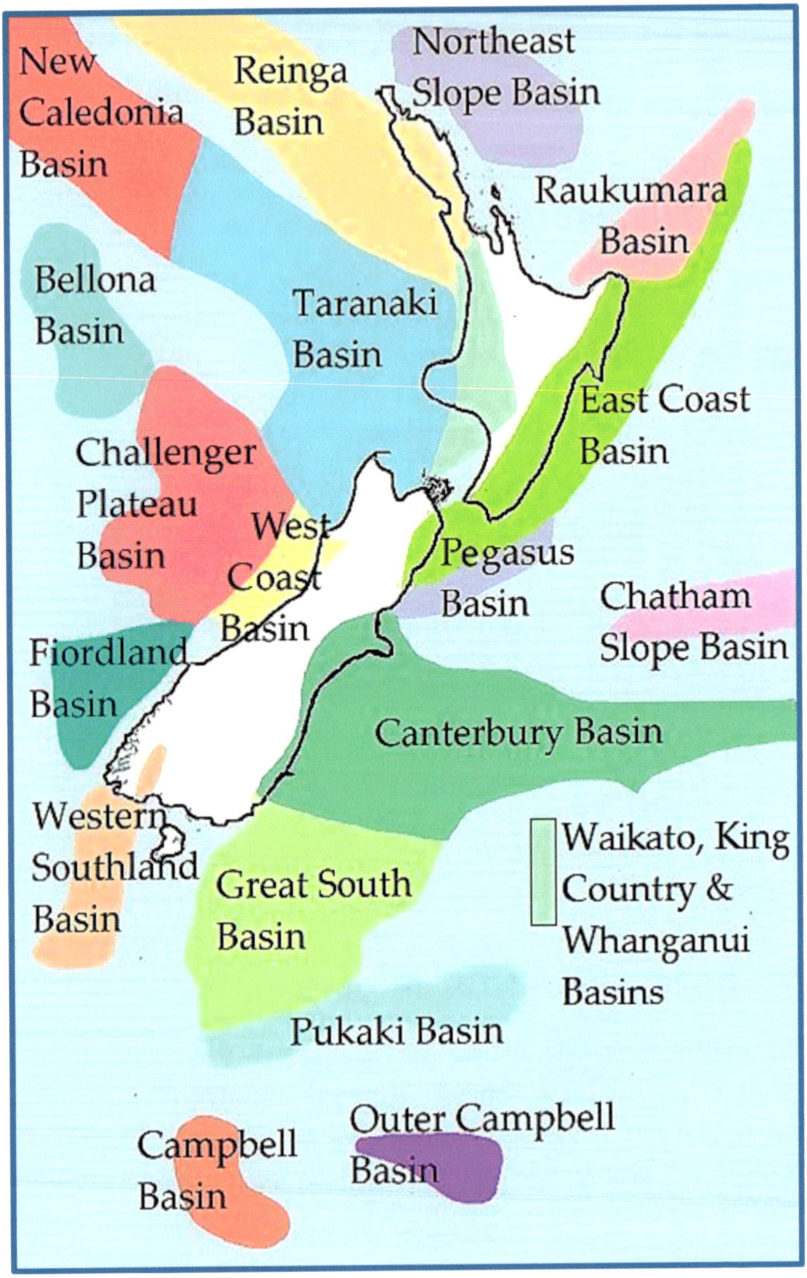
New Caledonia Basin
Reinga Basin
Northeast Slope Basin
Raukumara Basin
Bellona Basin
Taranaki Basin
East Coast Basin
Challenger Plateau Basin
West Coast Basin
Pegasus Basin
Chatham Slope Basin
Fiordland Basin
Canterbury Basin
Western Southland Basin
Great South Basin
Waikato, King Country & Whanganui Basins
Pukaki Basin
Campbell Basin
Outer Campbell Basin

The Taranaki Basin, an area of about 330,000 km², is mainly off-shore in production. Although under-explored with considerable potential for further discoveries, since 1950 there have been over 650 exploration and production wells. Those wells on-shore, approximately 450, are of a lower set-up cost and of lower production.

Fossil Fuel Basins	Km^2
Canterbury Basin	360,000
Taranaki Basin	330,000
North Caledonia Basin	300,000
Reinga Basin	170,000
Great South Basin	130,000
Challenger Plateau Basin	120,000
East Coast Basin	120,000
Northeast Slope Basin	80,000
Bellona Basin	80,000
Pukaki Basin	60,000
Campbell Basin	40,000
Waikato, King Country, & Whanganui Basins	40,000
Western Southland Basin	40,000
Chatham Slope Basin	40,000
Raukumara Basin	36,000
Fiordland Basin	35,000
West Coast Basin	25,000
Pegasus Basin	25,000
Outer Campbell Basin	20,000

Chapter 5

Petroleum Fields

2P reserves (proven plus probable reserves) of petroleum in Taranaki Basin fields, as at January 2022:

Petroleum Fields in Taranaki Basin		Crude Oil MB	Gas Pj
Kauri	Off-shore	2.4	4
Kupe	Off-shore	6.3	214
Maari	Off-shore	3.7	3
Māui B	Off-shore	9.6	314
Pohokura	Off-shore	6.5	342
Cheal	On-shore	1.2	2
Kapuni	On-shore	4.4	166
Kowhai	On-shore	0.2	10
McKee	On-shore	0.2	35
Mangahewa	On-shore	9.2	430
Ngatoro	On-shore	1.4	18
Radnor	On-shore	0.2	6
Turangi	On-shore	13.9	452
Waihapa-Ngaere	On-shore	0.4	2
Total		**59.6**	**1,998**
2021 Production		**7.6**	**157.3**
2P production years left at current production rate		**7.8**	**12.7**

Note 1:

In addition to the 2P reserves from these 14 fields, there are a number of smaller producing petroleum fields.

Note 2:

MMBBl (MB in the table) is an American acronym, the number of oil barrels. The word, 'barrel', is abbreviated to BBl, the first 'B' is thought by some to be as when the barrels were introduced they were of a blue colour (and now as seen below).

If one 'M' as in MBBl, the number is in thousands of barrels, say 5,000 barrels would be written 5MBBl. If two M's as in MMBBl, the number of barrels is in millions, 5,000,000 barrels would be written 5MMBBl, or otherwise abbreviated to 5TB and 5MB!

Blue Barrels

Measurement of crude oil in New Zealand is in barrels or litres. USA measurement of crude oil began as 42 gallons per barrel (or blue barrels, BBl), that is, almost 159 litres per barrel. In Europe, because of transporting via sea tankers, the measurement in tons became the norm.

Although conversion is imprecise due to weight differences from impurities in oil, the BP Statistical Review of World Energy has established conversion for crude oil at 7.33 barrels per metric tonne (1165 litres of crude oil per tonne).

That is, if crude oil is traded by:

- the barrel, about 159 litres of crude oil is transferred,
- the tonne, about 1165 litres of crude oil.

Crude oil yields over 6% refinery processing gain due to a lower density of the products processed than the raw crude oil itself (by the removal in crude oil of sulfur and salt), that is, increase to about 170 litres of refined products per crude oil barrel, or 1,240 litres per tonne of refined products.

Note 3:

Petajoules (Pj) is the energy produced. Converting barrels of crude oil into Pj, the above total of 59.6 million barrels of 2P crude oil reserves is equivalent to **365 Pj** (1MB of crude oil is equivalent to 6.12 Pj) — compared to, as in the above table, for 2P of gas, nearly **2,000 Pj**.

Note 4:

The abbreviations, 1P is proven reserves, 2P the sum of proven and probable reserves (at least 50% commercially and technologically recoverable), and 3P the sum of proven, probable and possible reserves. The best estimate of recoverable reserves is 2P.

Appraisals are made of fields to calculate their contingent resources, that is, resources potentially recoverable depending on future commercial and technological viability.

Abbreviations for contingency fields are high, best and low, 1C, 2C and 3C respectively. The largest 2C crude oil and gas fields in the Taranaki Basin:

Fields	Crude Oil MB	Crude Oil Pj	Gas Pj
Rimu	55	340	
Kauri	43	260	
McKee	33	200	
Mangahewa			1,200
Kapuni			700

Crude oil is converted to Pj for comparison to gas.

Petroleum reservoirs may be classified from their change of state between gases and liquids.

Reservoir Type	Description
Dry gas reservoirs (natural gas, page 45)	Gaseous that remains gaseous when at ground surface.
Wet gas reservoirs (gas-condensate, page 48)	Primarily gaseous, although contains a small amount of liquid crude oil that has condensed.
Volatile oil reservoirs (petrol, page 52)	Very light oil, although may vaporise at the surface into gas.
Black oil reservoirs (bitumen, page 63)	Heavier oil that is 'dead', has high viscosity and does not vaporise.

Chapter 6

On-shore Production

Crude oil and gas in reservoir sedimentary rock is located by seismic reflection and refraction. A small explosion is set off and low-frequency shock waves travel rapidly underground, bouncing off different types of rock, the waves returning to the surface. Geophones interpret these sound waves to determine the underground geological layout and the possibility of a petroleum reservoir.

If a potential crude oil or gas reservoir is located, a rotary drilling rig with a derrick structure is set-up (depicted overleaf). A wire rope from a winch drum on the rig floor runs over the crown block at the top of the derrick and down to the travelling block to lower drill pipes (known as 'drill string') and steel casings down into the well bore.

Attached at the end of the lowest drill pipe is a hollow, circular steel drill cutting bit with tungsten carbide teeth, or if harder rock is encountered, diamond teeth, to cut a bore through earth and rocks.

The casing is inserted into the well bore to prevent the risk of contaminating the surrounding aquifers from bore drilling 'muds' and 'fluids' (info on these overleaf). Between the casing and the rock wall is pumped cement for strength and stability of the casing, with the 'drill string' inside the casing.

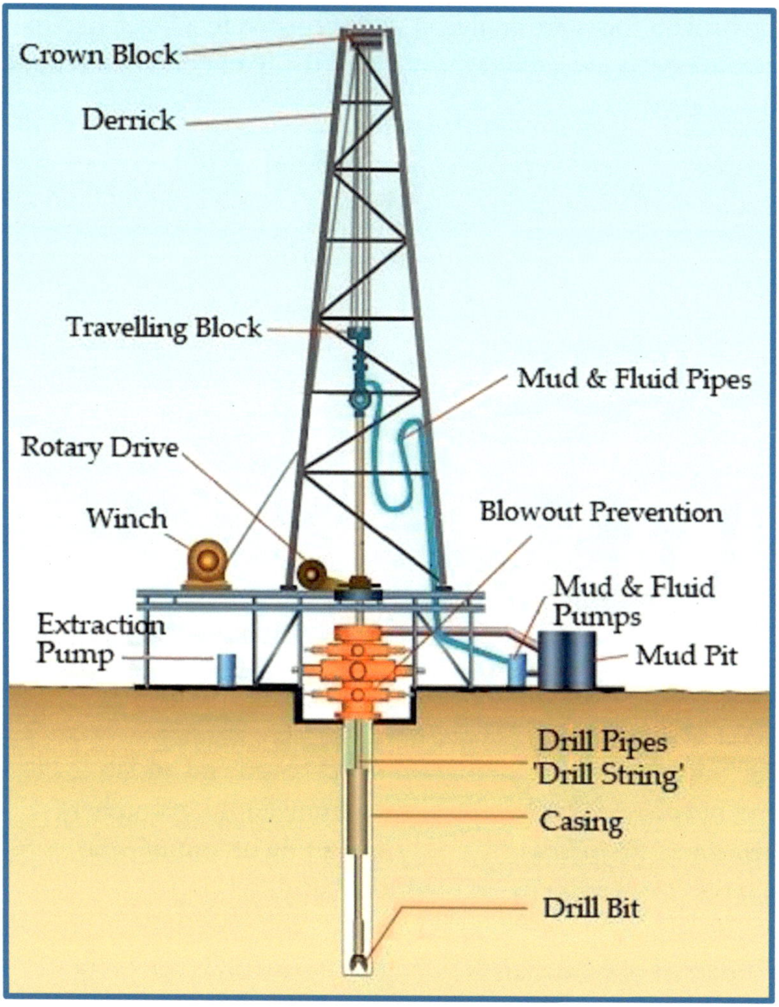

Drilling 'muds' are used to cool and lubricate the drill bit, and acts as medium to carry back up to the rig, between the 'drill string' and the casing, known as the annular space, 'cuttings', small pieces of broken rock. The 'cuttings' are forced up to the rig by the aid of 'fluids', compressed gas, pumped down the centre of the 'drill string'.

The broken rocks are analysed and recorded in a 'mud log' to determine the geology of the site and the likelihood of a crude oil and natural gas reservoir.

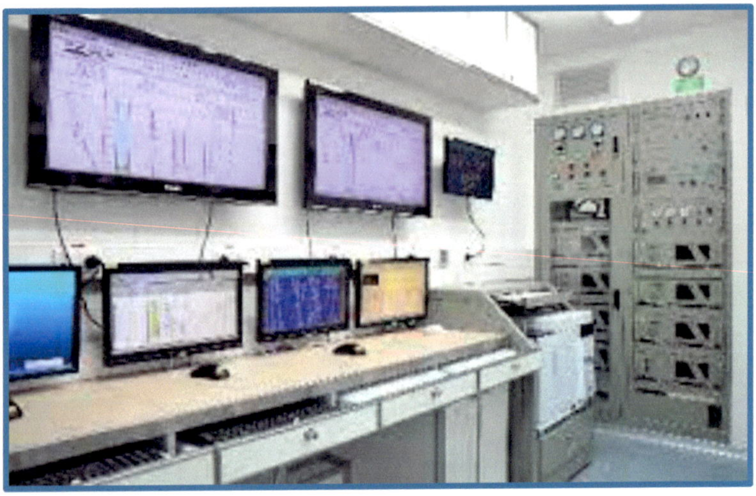

Mud Log Office

The 'muds' may be water-based or oil-based and added to the mud may be emulsifiers (to suspend one liquid in another), surfactants (to increase the easy spreading of 'muds') and gellants (to increase the viscosity of 'muds').

If suitable source rock is located, the bore drill is remotely controlled to turn horizontal in lateral directions to travel up to a few kilometres, allowing access of potential oil and gas across a larger area rather than just beneath the rig.

Horizontal drilling minimises land disturbances with less rigs and bore holes.

Chapter 7

Off-shore Production

Off-shore drilling for crude oil and gas in New Zealand began in the 1960's, and now over 200 wells have been drilled, about 10 in deep water of over 300 metres to the seabed.

Off-shore seismic surveying involves a survey vessel that shoots out seismic waves into seabed rocks, reflecting back to hydrophones for analysis. The data collected shows folds and faults in the seabed, which often act as traps for crude oil and gas that have moved upwards through source rock.

Generally, the same drilling techniques are employed as in on-shore rigs.

The difficulty in deep water crude oil and gas production is:

- At that depth, the water is near freezing, uncomfortable for diving teams.
- As such, equipment needs to be installed, maintained and operated remotely from the drilling rig.
- At that depth there is greater pressure on equipment, requiring it to be strongly constructed.
- It requires riser pipes that connect the well bore at the seabed with the drilling platform, for the circulation of drilling 'muds' and 'fluids'.
- Weather extremes are more severe further from the shore, with difficulty in dispatching necessary supplies.

Spills occasionally occur, the release of oil into the ocean as a consequence of equipment failure or of mis-management. A spill occurred from the Umuroa, an oil production and storage vessel in the Tui Field, 2007, which spilled about 23 tonnes of oil into the ocean (about 170 barrels of oil).

Off-shore Drilling Platforms

May be massive and contain accommodation and a helicopter landing pad. Drilling rigs may be designed for later conversion to production rigs, reducing the need to change rigs, if oil is located.

Drilling platforms may be:

- **Barge rigs** in shallower waters, usually less than 5 metres in depth. The barge hull rests on the seabed for stability.
- **Submersible rigs** in waters up to 15 metres in depth (depicted next page). A submerged barge with the drilling unit rests on the seabed, and on the barge's deck are steel posts that extend above water-level to the drilling platform with rig. These rigs are commonly used in areas of calm water.
- **Jack-up rigs** in waters of up to 150 metres in depth, the most common type of rig. The drilling platform with rig and drilling unit is supported on extendable steel posts anchored to the seabed.

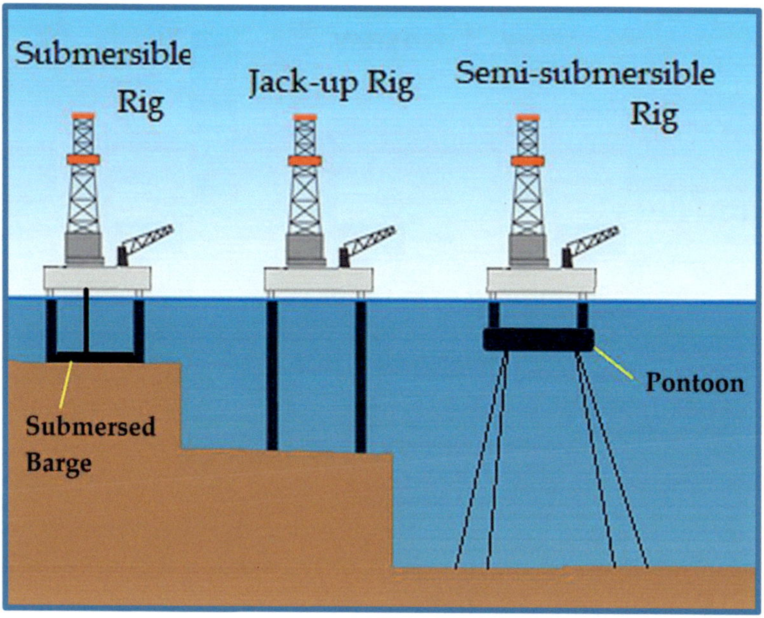

- **Semi-submersible rigs** in deeper waters. The drilling platform with rig and drilling unit is supported by steel posts to an underwater ballasted pontoon with sufficient buoyancy. The pontoon, anchored by up to a dozen chains to the seabed, often contains computer controlled dynamic positioning thrusters. The underwater pontoon helps to stabilise the platform as the pontoon is below the sea surface in calmer waters.

- **Drilling vessels** in deep waters, with the drilling rig and unit on the vessel's deck, rely on anchors and computer controlled dynamic positioning thrusters to maintain its position over the well.

Chapter 8

Blow-outs

If the bore drill locates oil or gas with an immediate release under pressure, this will result in a 'kick', and if not controlled, a blow-out, a geyser or a 'gusher', oil and gas and often water, drilling 'fluids', 'muds' and rocks shooting metres into the air.

The 'gusher' may ignite, resulting in injury, loss of equipment, and loss of oil and gas, as well as affecting the environment.

The Mangahewa Stream in Taranaki was contaminated for 18 months after the 1995 McKee well had a blow-out.

The line of defence is:

- **Drilling 'muds':**
  Prevents a 'kick' by a continuous counter-balance being maintained of a greater pressure of drilling 'muds', with sufficient density, pumped down the bore well than from pressure release of a 'kick' from oil and gas in the reservoir.

- **Hydraulic operated blow-out preventer:**
  A specialised valve that may stop a 'gusher' if this occurs. During drilling the preventer is left open to allow free-flowing of the drilling 'muds' and 'fluids', as well as the petroleum.

- **Capping stack:**
  An over 80 tonne 'capping stack' connected to the top of the blow-out preventer seals off the well if the hydraulic preventer fails to stop the 'gusher'.
- **Shear rams:**
  Are a last line of defence and are activated to cut through the 'drill string' to seal off the well completely.
- **Relief wells:**
  Drilled at depth below ground level into the 'gusher', to force in high density kill-weight muds, laden with barium sulfate ($BaSO_4$) to increase the weight of the mud, and thus create pressure resistance.

Maritime New Zealand has a range of equipment, plus an oil recovery vessel, to respond to an off-shore oil spill. The equipment is purchased and maintained through levies paid by production companies.

If a major spill occurs, specialist equipment may need to be called in from Australia or Singapore.

Chapter 9

Hydraulic Fracturing

Hydraulic fracturing (in short, fracking) may be used to extract crude oil and gas, where they have been trapped tightly within source rock, not being able to be expelled or to migrate.

A perforating gun is lowered into the bore well and fired into the reservoir sedimentary rock, creating fissures. The firing is gradually repeated along the horizontal lateral bore holes until all is fracked.

After fracking, high-pressure water with an added proppant is injected, for example with sand or ceramic beads, to keep the small fissures in the source rock open for the oil and gas to migrate to the bore holes.

Chemicals are necessarily added to the injected water to prevent bacterial build-up and corrosion, and for easier spreading of the proppant.

Fracking 'fluids' may be re-recovered and used further, for the wells to be re-fracked to extend their productive life.

Fracking may induce earthquakes, although of generally less than magnitude 2 on the Richter Scale.

Fracking may also cause water contamination by migration of oil, gas and chemicals through the created fissures, eventually reaching aquifers.

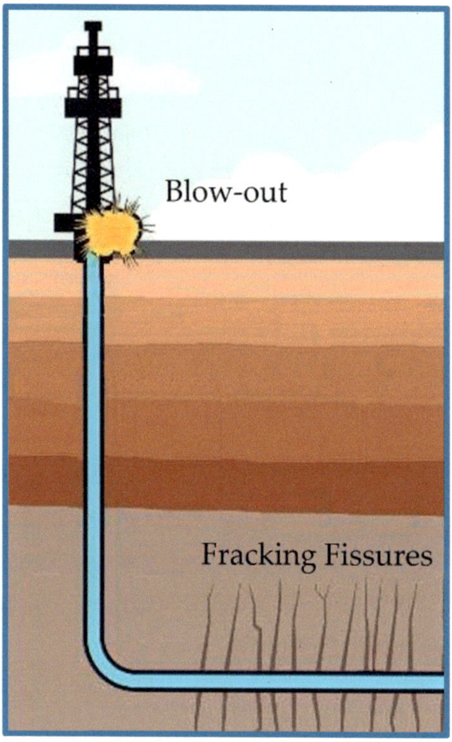

Chapter 10

'Cuttings'

Drilling for oil and gas produces 'cuttings', consisting of sedimentary rocks and earth, and both drilling 'muds' and fracking 'fluids' with additives, that need to be disposed of environmentally, preferably at the lowest cost.

'Cuttings' can be spread and then ploughed into farmland, left to fallow and periodically tilled to aerate the soil, which allows microbes to decompose the matter.

This process is referred to as 'land farming', in that less productive land, especially of sandy consistency, retains more structure with 'cuttings', and thus enables an increase in land productivity. The higher the mineral content, the longer the cuttings take to degrade.

Land farming — ploughing 'cuttings' into pasture land

Chapter 11

Royalties

New Zealand Petroleum & Minerals (NZP&M) regulates the issuing of prospecting, exploration and production permits to New Zealand and foreign petroleum companies. The Crown receives royalties from the companies for the production and sale thereof of minerals:

- Higher valued resources, such as petroleum, coal, gold and silver, are subject to royalties on a percentage of net sales or on company profits, whichever is the higher.

  Royalties on petroleum products are regulated by the Crown Minerals (Royalties for Petroleum) Regulations 2013.

  The royalty for oil and gas is presently 5% of the net sale revenue or 20% of accounting profits, whichever is the higher.

- Lower valued resources, such as road metal, are subject to a royalty based upon per tonne extracted and sold.

Chapter 12

Crude Oil & Gas Composition

	Fuel Type	Hydrocarbon Composition	Hydrocarbon Name
Gasses	Natural gas Liquefied natural gas Compressed natural gas	C_1H_4 C_2H_6	Methane Ethane
	Liquid petroleum gas Gas-condensate	C_3H_8 C_4H_{10}	Propane Butane
Liquids - Oils	Gas-condensate	C_5H_{12}	Pentane
	Petrol	C_5H_{12} C_6H_{14} C_7H_{16} C_8H_{18} C_9H_{20}	Pentane Hexane Heptane Octane Nonane
	Bio-ethanol	Petrol + fermented sugars or starches	
	Aviation fuel	C_9 - C_{16}	
	Diesel	C_{12} - C_{20}	
	Bio-diesel	Diesel + vegetable oils or tallow	
Semi-Solids	Fuel oil for ships	C_{20} - C_{70}	
	Waxes	C_{20} - C_{40}	
	Bitumen for asphalt and chipseal road seal	C_{35+}	

Crude oil may be light crude — yellowish to nearly clear, to heavy crude — dark brown or black, producing various products as in the previous table, depending upon their hydrocarbon compositions, the number of carbon (C) and hydrogen (H) atoms in each molecule.

As the number of carbon atoms increase, fossil fuel is formed into, from gases, to liquids (light oils through to fuel oil for ocean ships) and then to semi-solids (waxes through to bitumen for both asphalt and chipseal road seal).

The boiling and melting point also tend to rise as the number of carbon atoms increase, and the hydrocarbon molecules become heavier and larger, and darker and denser, and with greater viscosity.

API Gravity

The American Petroleum Institute established the API gravity, an inverse measure of petroleum's specific gravity or density compared to water. API is measured in degrees on a hydrometer.

Approximate API of gases and crude oils:

Product	API
Gases	
Natural Gas	75 - 90°
Gas-condensate	45 - 85°
Crude Oils	
Light oil	30 - 45°
Medium oil	20 - 30°
Heavy oil	10 - 20°
Extra heavy oil	Below 10°

Water has an API of 10°. If petroleum has an API greater than 10°, the petroleum floats on water, and if less than 10° sinks in water, 'extra heavy crude oil'. The lightest oil is about half as heavy as water.

Crude oil floats on water as it has less density than water, usually 80 to 90% to that of water.

Oil floats on water after an oil tanker breaks up

Petroleum of API between 35 and 45° is of the greatest value. Most crude oil in the Taranaki Basin is 'sweet' (contains little sulfur), waxy and about 40 to 45° API.

By assessing the weight of crude oil by its API gravity, this enables the number of barrels to be calculated per tonne of crude oil, and its market value per TB or MB.

Chapter 13

Sulfur & Salt in Crude Oil

The higher the amount of impurities in crude oil, the lower the market value. The most common impurities are sulfur and salt. Removal of sulfur and salt in crude oil increases the API, with a refinery processing gain of litres per tonne of refined petroleum.

Sulfur

Crude oil is 'sour' that has more than 0.5% by weight of sulfur (S), and is 'sweet' with a lesser amount.

Sulfur may also contain hydrogen sulfide gas (H_2S). Sulfur-reducing anaerobic bacteria, which use sulfur as an energy source, is the primary cause of hydrogen sulfide.

When natural gas is produced as a by-product of crude oil, to remove, if present, the highly toxic hydrogen sulfide gas, the natural gas along with the hydrogen sulfide may be 'flared'.

Flaring natural gas at Mangahewa well, McKee **field**

Sulfur that may be in petroleum needs to be removed, as:

- Sulfur competes with other pollutants for space on the vehicle's exhaust catalytic converter surface, which may inhibit removal of other pollutants (refer Chapter 20 for catalytic converters).
- During fuel combustion, the sulfur oxidises to sulfur dioxide gas (SO_2).
- Sulfur is corrosive to vehicle parts.

Removal of sulfur in crude oil may be by:

- Conversion, the 'cracking' of hydrocarbon molecules to release sulfur.
- Treating with bentonite, an absorbent swelling clay.
- Treating with an alkaline solution, sodium plumbite (Na_2O_3Pb).
- Hydro-treating, the reaction by catalyst, copper, in the presence of hydrogen (H), at relatively high temperature and moderate pressure, converting sulfur to hydrogen sulfide gas and water vapour (H_2O).

Hydrogen sulfide gas,

- which may be with natural gas when extracted from reservoirs, or
- which is produced from hydro-treating to remove sulfur from petrol

oxidises in the atmosphere to form sulfur dioxide gas.

Combustion of sulfur in vehicles also produces sulfur dioxide gas.

Sulfur dioxide is neutralised in the air by a spray of 2% sodium bicarbonate.

When sulfur dioxide is dissolved in rain, sulfuric acid droplets (H_2SO_4) is produced, which causes smog and 'acid rain'.

The reaction of dissolving phosphate rock in sulfuric acid produces phosphate fertiliser, the main product use of sulfur when removed from fossil fuels.

Sulfur dioxide and hydrogen sulfide gas prevents cellular respiration, and causes irritation to the eyes, convulsions and dizziness. Sulfuric acid causes severe skin burns and difficulty in breathing if inhaled.

Salt

Crude oil often contains salt crystals, mainly sodium chloride ($NaCl$), although also calcium chloride ($CaCl_2$) and magnesium chloride ($MgCl_2$), which can corrode piping, and thus requires removal before refining of the crude oil.

The crude oil is heated up to 150°c and then mixed with up to 10% water to dilute the salt. After this mixture settles, the crude oil, as lighter than water, is drawn off.

Chapter 14

Refinery Production

Refinery production is the process of crude oil fractional distillation, the separation into fractions of various energy products – petrol, diesel, aviation fuel, fuel oil for ships, liquid petroleum gas (LPG) and compressed natural gas (CNG), plus petrochemical products, all with their specific hydrocarbon molecule consistency.

The distillation process occurs in a hollow carbon steel tower that is typically 30 metres in height and 6 metres in diameter above its base, although can be twice the height and twice the diameter. Inside the distillation tower are up to 40 staggered distillation plates for collection of the various petroleum products, by the process of vaporisation and then condensation of the crude oil.

The crude oil is heated by a fuel-fired reboiler (a heat exchange unit to generate steam in the distillation tower), to over 400°c for vaporisation of the oil, the vapour entering the bottom of the distillation tower.

Crude oil tar, 'residue', that is too dense to vaporise, drops to the bottom of the tower for further processing in a vacuum distillation tower under reduced pressure, enabling boiling and separation from impurities at a lower temperature, to extract from the tar, heavy coke fuel for steel manufacturing and bitumen for roading.

The diagram below shows the approximate temperature that the hydrocarbon molecules condense (vapour changes to a liquid), with their number of carbon atoms, to produce the various 'distillates'.

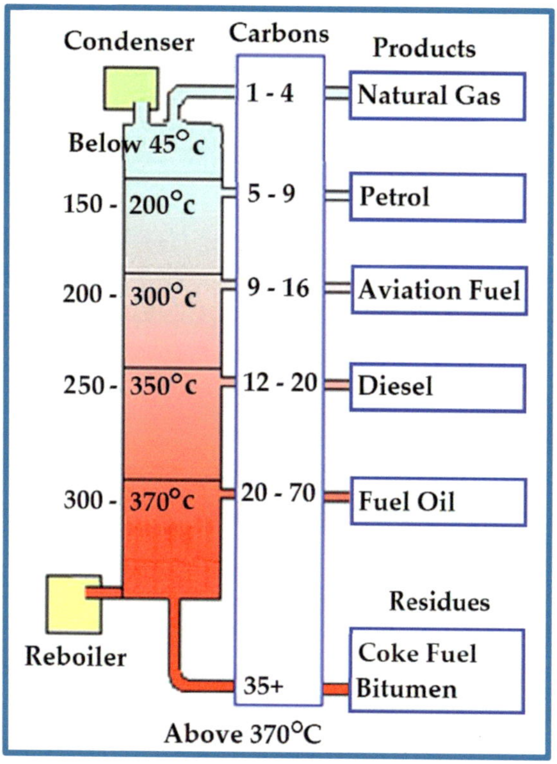

As the vapour in the tower rises, the vapour condenses as it cools. At the first distillation plates, a temperature of about 300°c, the heavier hydrocarbon molecules condense at their specific boiling point to collect on the plates as 'distillates', fuel oil for ocean ships, to be drawn off, then sent to either a storage tank or for further processing.

As the vapour continues to rise, it cools, when at about 250°c the next 'distillate', diesel begins to emerge, at 200°c aviation fuel, at 150°c petrol, and at the top of the tower, 45°c, natural gas remains.

A condenser, a heat exchange unit near the top of the tower, cools and condenses emerging water vapour into a liquid 'stream', refluxing this down to the reboiler at the base of the tower, to assist in further vaporising the incoming crude oil for distillation in the tower.

At the bottom of the tower, the 'distillates' produced are of large molecules with a high boiling point and viscosity, and do not ignite easily (high flash point).

At the top of the tower, the 'distillates' produced are of small molecules with a low boiling point, and are very volatile (change between liquid and gas) and flow and ignite easily (low flash point).

As fractional distillation produces excess 'distillates' of a higher carbon consistency, and lesser lighter 'distillates', petrol and diesel of greater demand, the process, 'cracking', breaks down those larger crude oil hydrocarbon molecules into products of smaller molecules, for petrol and diesel.

'Cracking' may be by thermal or catalytic action.

Thermal 'cracking' is the heating of heavier 'distillates', residue coke fuel and bitumen, to a temperature of over 500°c and with a high pressure, breaking down the complex hydrocarbon molecules.

Catalytic 'cracking' uses catalyst, zeolite, at moderate temperature and pressure, which causes a chemical reaction to break the hydrocarbons into smaller molecules, products of petrol and diesel.

Catalytic Cracker

Also, lighter hydrocarbon molecules may be combined to form petrol, diesel and other products (unification), or by altering the structure of hydrocarbons although with the same number of carbon atoms (alteration).

Chapter 15 Fuels used in NZ Transport

Fuel consumed	Petajoules	Percentage
Electric	0.4	0.2%
Hydrogen gas	0	0
Regular petrol	60	30%
Premium petrol	25	12%
Diesel	100	50%
Aviation fuel	12	6%
Fuel oil	3	< 2%
LPG	< 0.1	0
CNG	< 0.1	0
Total	**200**	**100%**

Fuel	Used in vehicles
Electric	Passenger vehicles, buses, trains
Hydrogen gas	Passenger vehicles, buses, trucks
Petrol & Bio-ethanol	Passenger vehicles, light trucks, buses
Diesel & Bio-diesel	Passenger vehicles, trucks, buses, trains, fishing vessels
Jet fuel	Turbine jet engines
Avgas	Piston engine aircraft, tractors
Fuel oil	Ocean ships
LPG	Passenger vehicles, forklifts, hot air balloons
CNG	Passenger vehicles, buses
Methanol & Nitrous oxide	Racing cars

Chapter 16

Gas Fuels

Natural Gas

Natural gas is mostly methane and with some ethane. Methane is the smallest hydrocarbon molecule and has one carbon atom and four hydrogen atoms, C_1H_4. Ethane has two carbon and six hydrogen atoms, C_2H_6 (as the number of carbon atoms in a petroleum molecule increase by one, the number of hydrogen atoms increases by two).

Both the on-shore Kapuni gas field, commissioned in 1970, and the off-shore Māui A gas field in 1979, enabled New Zealand to be self-sufficient in gas. However, the Māui A gas field is now approaching depletion, although Māui B gas field has the 4th largest 2P reserve, 314 Pj of energy.

The gas market in New Zealand is independent to the rest of the world (does not usually import or export gas), and during 2021 natural gas accounted for 25% of indigenous energy production and 12% of consumer energy in New Zealand.

First Gas Limited reticulates natural gas around the North Island, compressed at a pressure of mostly 210 to 420 kilopascals (kPa), then delivered to consumers through lower pressure, local distribution networks. Natural gas is not reticulated in the South Island.

Natural gas is used in the petrochemical sector, in industrial processing, for heating of buildings and heating of water, and for cooking.

Ethanethiol, an odorant, may be added to give gases the smell of rotten cabbage so that any leaks may be detected.

Liquefied Natural Gas

Natural gas is liquefied at super-cool temperatures, below -160°c, which then has a volume 1/600th of its gaseous form.

Used as transport fuel, LNG evaporates when vehicles are left idle for any length of time. Because of evaporation and the cost of super-cooling of natural gas, it is not viable to liquefy in New Zealand as a transport fuel.

LNG, compared to diesel, has about 75% less nitrogen oxide (NO_x) and up to 15% less carbon dioxide (CO_2) emissions, as well as no particulate matter (PM_{10}) emissions. For this reason, and for less idle time in gas loss, LNG is a more suited fuel for long-distance truck haulage in Australia, America and Europe.

Compressed Natural Gas

CNG is natural gas at a pressure in cylinders not exceeding 20 MPa, and which is compressed to 1/100th of its gaseous form.

CNG was a government-subsidised fuel with low-interest rate loans for vehicle conversions using fitted cylinders during the 1980's when oil prices were high. CNG declined in use after the subsidies were removed in 1987 due to then, a drop in oil prices.

CNG is suitable for city buses, as necessity is shorter runs between fill-ups, and is environmentally friendly compared to petrol or diesel.

Liquid Petroleum Gas

LPG is denser than natural gas as it contains more carbon atoms per molecule, being usually a mixture of 60% propane (C_3H_8) and 40% butane (C_4H_{10}). However, as LPG is a lighter fuel than petrol and diesel, it has less harmful exhaust emissions than those fuels.

At an ambient temperature of 20°c, propane needs to be pressurised at 835 kPa to liquefy, and butane at 115 kPa. A mixture of these gases is between these two pressures. Also, the higher the temperature of gas vapour, propane and butane, a higher pressure is required to liquefy the vapour.

LPG is liquified and distributed around New Zealand by ocean or road tankers, and is also transported and sold in cylinders. The volume of liquefied LPG is around 1/270th of its gaseous form.

New Zealand LPG Cylinder Truck

New Zealand LPG Tanker

LPG is also shipped to the Pacific Islands.

Vaporised LPG is reticulated to many parts of New Zealand, especially the South Island where natural gas is not available.

LPG is used as fuel in over 10,000 motor vehicles, for heating of buildings and heating of water, for cooking, and in industrial processing plants.

Gas-condensate

Gas-condensate is condensation to a liquid state of natural gas, 'wet gas', the formation depending upon temperature and pressure. Condensate may be composed of propane (C_3H_8) to pentane (C_5H_{12}), and may naturally be formed in underground reservoirs or in pipelines.

Gas-condensate is hazardous, as the highly flammable vapours are heavier than air and so accumulate in low lying areas.

Condensate may be used to dilute viscous oils that cannot otherwise efficiently flow through pipelines.

Gas Hydrate

Gas hydrate ($C_4H_{18}O$) is formed from methane molecules in oceanic coarse-grained, sandy sediments of high pressure and freezing temperatures, located at the base of the 'continental slope', and also in permafrost areas of polar regions at about 200 metres in depth below sea-level.

The methane molecules are produced from decomposition of marine organic matter by the metabolism of archaea micro-organisms, the gas molecule becoming encased in a lattice of frozen water molecules.

New Zealand's largest gas hydrate reservoir is in the East Coast Basin. Difficulty in extraction is due to the volatility of the gas and the expansion as it rises to the surface, from a high to a lower pressure, with an approximate 165 times volume increase.

Research is on-going for production of gas hydrates, as methane has the lowest environmental impact of fossil fuels.

Hydrogen Gas

'Green hydrogen' has recently been developed to be able to power electric motor vehicles. Late 2021, a green hydrogen plant of 1.5MW production facility was commissioned by Halcyon Power in partnership with a Japanese corporation, located at Mokai, close to Taupo. Previously, hydrogen gas for energy was produced from fossil fuels.

FCEV re-fuelling station in Europe

Hydrogen gas (H_2) is odourless, colourless and non-pollutant.

Electricity, transformed from the Mokai geothermal power plant, splits water (H_2O) by electrolysis into hydrogen (H) and oxygen (O). Compressed hydrogen gas then enters electro-chemical fuel cells and transforms the chemical energy into electrical energy to power vehicles.

EVs (electric vehicles) use plug-in batteries to recharge; FCEVs (fuel cell electric vehicles) use green hydrogen gas for energy to generate their electricity on-board.

FCEVs are more efficient than combustion engines, and do not produce pollutant exhaust emissions, only water vapour and warm air. And only a small high-voltage battery is required, as electrical energy is not drawn from plug-in batteries as in EVs.

There is now approaching 30,000 vehicles on New Zealand roads that are EV's.

Five Hyundai hydrogen-powered heavy trucks are being imported as a working trial, partly funded by Energy Efficiency and Conservation Authority (EECA). The trucks have a range of about 400 kilometres, the approximate driving distance between Auckland and Palmerston North.

A network of green hydrogen re-fuelling stations is required to support the hydrogen-powered vehicles. And construction of New Zealand's first re-fuelling station was commissioned, May 2022, in Palmerston North. It is expected that a network of 24 hydrogen re-fuelling stations will be operational throughout the country within the next 4 to 5 years.

The first FCEV re-fuelling station in NZ, being build near the Palmerston North Airport. Photo taken 3rd November 2022.

Chapter 17

Oil Fuels

Petrol

Petrol is typically in the range of pentane (C_5H_{12}) to nonane (C_9H_{20}), that is, 5 to 9 carbon atoms per molecule.

Petrol is more volatile and flammable than diesel, and contains fewer carbon atoms per molecule.

Petrol is sold in two or three grades. The difference in the grades is that premium petrol, with a higher octane rating, is more suitable for high-performance engines, formulated to be harder for the petrol vapour to ignite, reducing the tendency of engine 'knock'.

- Premium petrol does not contain significantly higher energy content, increasing power or fuel economy.
- Premium petrol may be formulated with only minimal higher quality additives to increase engine cleanliness and wear.

Octane rating is the measure of resistance to knocking, the comparison of 'knock' tendency of fuel to that of the 'knock' from the blend of iso-octane fuel (which is resistant to knocking and given a rating of 100) with heptane (the mid-range of petrol, C_7H_{16}).

The octane rating is the percentage of iso-octane in the blend that matches the 'knock' in the fuel. In say, 98 octane petrol, 98% of the blend, iso-octane with heptane, would be iso-octane. Iso-octane has the same atoms as octane, (C_8H_{18}), although with different chemical properties.

To prevent confusion and fraud, petrol grades may be dyed different colours for distinction, although is not in New Zealand. Instead, regular testing is carried out for compliance and safety.

Motorists of high performance cars that operate on premium petrol may become aware of grade confusion or fraud only when the regular grade petrol in their tanks cause 'knocking' and their engines perform poorly, eventually with damage.

Colour distinction overseas is often:

- Regular – greenish to bluish
- Mid-grade – yellowish
- Premium – pinkish

Petrol at the pump is transparent-like, and as it ages the petrol oxidises with a sour scent and a brownish colour.

Such petrol is not of suitable use due to:

- Ignition problems
- Poor acceleration
- Eventually causing engine damage

Lead (Pb) **in Petrol**

In the 1920's, tetraethyl lead (TEL) was added to petrol as it raised the fuel's octane rating, resisting engine 'knock', and plus protecting the engine's pistons from excessive wear.

However, leaded fumes cause serious health risk, especially in children, lowering their IQ. Allowable lead in petrol was reduced in 1996, and now the maximum amount in petrol under the Engine Fuel Specifications Amendment 2011 (EFSA) is 5mg per litre.

Pre-1990 cars need to be assessed if able to operate on 'unleaded petrol', and if not, a fuel with suitable octane rating is required, or with a lead replacement additive.

Bio-ethanol fuels are unsuitable for engines designed to take 'leaded fuels'.

Methanol

Methanol (CH_3OH), also known as methyl alcohol, is colourless with a faintly sweet odour, and is bio-degradable as it breaks down in air and is soluble in water. Methanol occurs naturally in the environment.

Methanol may be produced from methane. Methane (CH_4), the main gas in natural gas, is 'steam reformed' to produce synthesis gas, commonly known as 'syngas', consisting primarily of carbon monoxide (CO) and hydrogen (H), which this, then by reaction from catalyst, copper or zinc, produces methanol. Methanol, in turn, by reaction from catalyst zeolite, may be blended with petrol.

The octane rating of methanol is around 115 and blending converted methanol with petrol upgrades regular petrol to premium. However, an additive is necessary for methanol blended fuels to provide protection against corrosion.

Methanol as a vehicle fuel blend has:

- A high oxygen content for a cleaner fuel combustion, the chemical reaction in which fuel oxidises.
- A low carbon to hydrogen ratio for an environmentally friendly fuel.
- No sulfur contamination, which can inhibit the catalytic converter in the removal of other pollutants.

Methanol (CH_3OH) has low energy density as it contains only one carbon atom, although as methanol also contains one oxygen atom, a faster combustion occurs than with petrol (C_5H_{12} to C_9H_{20}), with a corresponding increase in power.

EFSA states methanol may be blended with petrol to a maximum of 3% by volume.

Methanol may be blended to a higher percentage with petrol as fuel for racing cars, to produce more power, for less smoke if a car ignites from a crash (no loss of visibility to other drivers), and for flames able to be extinguished with water.

The Motunui synthetic methanol fuel plant, located on the north Taranaki bight, the first of its type in the world, was developed under the 'Think Big' policy of Prime Minister Robert Muldoon, commencing production in 1987.

The plant converted natural gas from Māui A gas field into methanol, then to petrol. When oil prices fell soon after plant inception in the late 1980's, the plant changed to producing methanol for export. As the Māui A gas field began to deplete, the Motunui plant was decommissioned.

The plant sat idle for five or so years until 1993, when ownership transferred to Methanex NZ, a Canadian based company. The company secured sufficient natural gas rights from Taranaki fields and produced up to 2.2 million tonnes of methanol per year, approximately 95% exported to the Asia-Pacific region.

One of Methanex NZ's three plants ceased operations early 2021, unable to secure sufficient future gas supplies. This has been exacerbated by the Crown Minerals (Petroleum) Amendment Act 2018, restricting exploration (refer page 104). Although the plant will re-commence operation should gas supply become available.

Motunui methanol fuel plant

Methanol is in demand as a clean energy source. Used as a transport fuel and for heating buildings, in pharmaceuticals and for production of solvents, antifreeze, synthetic fabrics in clothing, adhesives, paints and plywood resin, to name but a few applications.

Naphtha

Naphtha emerges in the distillation tower at a temperature anywhere less than kerosene (aviation fuel) and is 'reformed' with catalyst, usually platinum (Pt), for the product to be blended for high-grade petrol, the improving of combustion.

The word, 'Naphtha', is derived from 'naft', a Persian word for describing bubbling oil.

Naphtha is classed as 'light' if less than 6 carbon atoms per molecule or 'heavy' if over 6. Light naphtha is often volatile and flammable.

Light naphtha is used:

- In fire performances on stage, as the flame is brighter, cleaner and of a shorter burn.
- As a fuel for portable stoves, lanterns and cigarette lighter fluids.
- As a paint solvent and cleaning fluid.
- As an oil painting medium.
- In laundry soaps and in varnishes.

Heavy naphtha is used to dilute viscous oils for moving more easily through pipelines.

Naphtha is carcinogenic and causes skin irritation.

Aviation Fuel

Aviation fuel is a kerosene type fuel, composed of carbons mainly in the range C_9 to C_{16}. The fuel is required to stay liquid at extremely low, high altitude temperatures.

Kerosene has a higher octane rating with a higher flash point, with greater power and efficiency than petrol. The flash point for kerosene is over 38°c and for petrol minus -43°c.

Flash point is the temperature at which the fuel gives off sufficient vapour to ignite in the air. Another reason for using a kerosene type fuel for aviation is safety as regards the flash point.

Aviation fuel is either jet fuel for turbine jet engines, or avgas fuel suitable for piston engine aircraft, which are mostly used in flying clubs, for agriculture aircraft and for private domestic flights.

Avgas is also suitable for the modern tractor, providing greater power when required.

Recently developed is blended sustainable aviation fuel (SAF), produced from, such as plant oil or animal fat. SAF is similar to conventional fossil jet fuel in its chemical composition, although emits up to 80% less carbon. At present, the cost of SAF is an issue, as it is up 10 times more costly to produce than traditional aviation fuel.

Diesel Fuel

Diesel fuel contains carbons in the range of C_{12} to C_{20}.

The name, 'diesel', is an eponym of Rudolph Diesel, a German scientist who, in 1894, invented the first engine that was fuelled with diesel.

As the production of diesel fuel requires less refining than petrol, retail prices are lower. Over 600,000 light vehicles in New Zealand, under 3.5 tonnes, run on diesel, and almost the entire truck and bus fleet, plus trains and fishing vessels.

A winter grade diesel is produced for the south, as when lower temperatures occur, ice crystals form in diesel, blocking filters and the engine's fuel supply.

In warmer temperatures, 'diesel bug', microbial colonies of bacteria or fungus living in water at the base of vehicle fuel tanks, feed on sulphur in diesel, multiplying rapidly.

Problems with 'diesel bug' include difficult engine starting and poor performance, as well as clogging of filters, fuel lines and pump. Biocides may eradicate 'diesel bug'.

Bio-diesel is more prone to 'diesel bug' as the bacteria and fungus also assimilate plant oils and animal fats in bio-diesel.

Fuel oil

Fuel oil, also known as marine, furnace or bunker oil, is the most viscous and heaviest of liquid fuels, condensing at between 300 and 370°c and with a carbon content C_{20} to C_{70}.

Fuel oil is any heavy petroleum fired in industrial furnaces, for the heating of large buildings, or for the powering of ocean ships.

The worldwide shipping regulator, International Maritime Organization for fuel oil, opted out of the Paris Climate Agreement, enabling ocean ships to continue using fuel oil, which produces a relatively higher amount of pollutants. The continuing use of fuel oil is partly due to the approximate 30% lower cost than of alternative fuels.

Bio-fuels

Bio-fuel is from bio-mass – the decomposition of plant or animal material, and is 'carbon neutral' apart from emissions in producing and transporting of the bio-fuels.

The bio-mass is 'renewable' and bio-degradable, and has no greenhouse and other pollutant gas emissions when used in vehicles and other industrial engines.

Bio-fuels are not dependent on new fuel infrastructures and required vehicle modifications, as with electric and hydrogen gas fuelled vehicles.

Bio-fuels are classed as either 'blended' (required to prevent engine damage), or recently developed, 'drop-in' (where the bio-fuel is chemically similar to fossil fuels).

Bio-fuel usage in New Zealand is less than 0.1%, compared on average with the rest of the world, at about 4%. And the use of bio-fuels in the transport industry has recently declined due to the increase of production costs in blending.

Bio-ethanol

A petrol fuel. The bio-mass is hydrous-ethanol, about 95% ethanol and 5% water. Ethanol (C_2H_5OH) is an alcohol type fuel produced from fermented sugars and starches, such as from maize or wheat, or from dairy whey, a by-product of cheese processing.

As ethanol is chemically more of a solvent than a petrol fuel, it can be more aggressive to motor parts.

The maximum hydrous-ethanol in petrol under EFSA is 10% by volume. However, most modern vehicles can take up to 25% hydrous-ethanol without modification.

Fuel with hydrous-ethanol of 10% is marketed as E10 and results in around 30% reduction in potent gas emissions compared to emissions from regular petrol. A more efficient combustion is from the higher oxygen content, and as a consequence bio-ethanol blended petrol has less pollutant emissions.

After timing and electronic control system adjustments, vehicles can operate on pure hydrous-ethanol (E100).

In cooler climates, as E100 may be harder to vaporise, a blend with petrol improves ignition (E85 — an 85% hydrous-ethanol blend).

Ethanol contains less energy output than petrol, although as ethanol has a higher octane rating, the engine can be made more efficient by raising the compression ratio.

Bio-diesel

The bio-mass in bio-diesel is produced from:

- Plant oils
- Tallow — animal fats
- Waste cooking oils — including inedible or recycled oils not fit for human consumption

EFSA states the bio-mass in diesel may be up to 7% by volume (marketed as B7), and can be used in diesel engines without modification.

Although the energy content of bio-diesel is less than straight diesel, bio-diesel has a higher oxygen content and thus has a more efficient combustion than straight diesel, reducing pollutant emissions.

Bio-diesel has a higher lubricity (slipperiness) than straight diesel due to the oil content in the bio-mass, reducing possible engine wear.

Bio-diesel tends to congeal in colder climates. Also, bio-diesel is more likely to gradually oxidise if the vehicle is left idle for a length of time.

In 2020, Z Energy produced 1.9 million litres of bio-diesel, although soon after the plant closed due to rising international tallow prices and falling diesel prices.

Bitumen

Bitumen is a viscous, heavy black product from the further distillation of crude oil tar in a vacuum tower under reduced pressure, enabling boiling and separation from impurities at a lower temperature.

Crude oil tar is commonly known as tar sand as the tar is often mixed with sand and clay.

Bitumen may also be extracted from crude oil tar by mixing the tar with hot water, enabling separation of the bitumen from the sand and clay.

Bitumen consists of high molecular weight hydrocarbons (the composition approximately 87% carbon, 7% hydrogen, 5% sulfur and 1% nitrogen).

Bitumen may be mixed with lighter crude oil to produce 'synthetic crude' oil. The word 'synthetic' distinguishes upgraded bitumen to form this fluid crude oil from that of conventional crude oil, the two products being chemically identical.

Bitumen has waterproofing and binding properties and is resistant to damage, thus suitable in roofing tiles and sheets, and as a binder with aggregate for road surfaces.

Coal tar is used to treat skin disorders, and is in shampoos.

Types of roads seals:

Chipseal seal, also known as bitumen seal

A bitumen layer is sprayed onto the road and then covered with aggregate. Machine rolling of the aggregate, binds the aggregate into the bitumen. This is then repeated to give a two or three-coat seal. Chipseal roads have more surface grip.

Chipseal - Higgins bitumen sprayer

Asphalt seal

Bitumen, aggregate and sand is mixed off-site, and then applied by a road paver machine to the road. Asphalt roads are smoother and more durable than bitumen roads, and vehicle noise is less.

Asphalt seal - Higgins road paver

Tarmac seal

Roads were previously tarmac, made direct from tar sand and aggregate, not of bitumen extracted from tar sand in a distillation tower. Because of impurities in tar sand the tarmac roads were less durable.

Coal tar seal

Coal tar, produced as a by-product of bituminous coal gasification (a process by which coal was converted into gas for cooking and heating) was used as a road seal prior to the 1980's.

Hydrocarbons from coal tar on the roads washed into roadside storm-water run-offs and then contaminated streams. It was previously thought that contamination of streams near roads was caused by pollutants emitted from vehicle exhausts.

Although roads are now chipseal or asphalt seal, they may still contain coal tar in their lower layers as roads are re-sealed several times before being rebuilt from their base.

Shale Oil

Oil shale is an organic-rich, fine-grained, laminated sedimentary rock, consisting mainly of silt and clay particles.

Through heat, of up to 50°c from decomposition of marine organic matter by anaerobic bacteria, and by the pressure of sediment above, the organic matter within the rock was transformed into a waxy substance, kerogen.

Lack of sufficient further heat and pressure has not yielded oil from the kerogen. Thus, to obtain shale oil, extracted oil shale is crushed and then heated in a reactor, a 'retort', at a temperature of between 450 and 500°c, causing the chemical process, pyrolysis, to decompose kerogen to either shale oil or a vapour that when condensed yields shale oil.

Oil shale can be burned directly in furnaces as a low-grade fuel.

As oil shale is high in carbon content and has about 50% more greenhouse gas emissions than crude oil, only when the price of crude oil is high, shale oil is considered as a choice to be processed into fuel products.

Around the year 1900, shale oil was extracted from fields in Otago and Southland, and recently the East Coast Basin has indicated economic exploration potential, where fields are up to 600 metres thick.

Chapter 18

Coal Fuels

New Zealand coal fields were formed from relatively recent geological activity, between 20 and 70 million years ago, after the continent of Zealandia broke away from the Australian continent.

Coal originates from peat, which is not fully decomposed plant and animal material.

The table below shows the progression of peat to the highest rank coal, anthracite, a hard lustrous coal with few impurities and of high energy density. Anthracite is not formed in New Zealand.

Coal Type	Burial Depth	Burial Temp.	Moisture Content
Peat	< 0.2 km	< 25°c	50 – 80%
Lignite	0.2 – 1.5 km	25 – 45°c	30 – 50%
Sub-Bituminous	1.5 – 2.5 km	45 – 75°c	10 – 30%
Bituminous	2.5 – 6.0 km	75 – 80°c	5 – 10%
Anthracite	> 6.0 km	>180°c	< 5%

As in the table, with an increase of sediment accumulation above peat, both temperature progressively increases and moisture content decreases, to form higher ranks of coal, from lignite to sub-bituminous, bituminous and to the highest rank, anthracite.

Peat

Contains elements, carbon (C), nitrogen (N), oxygen (O), hydrogen (H) and sulfur (S), plus organic material and ash. Peat is soft, and is easily compressed by accumulation of sediment above, which the peat may then form into lignite.

Un-drained peatland

Peat formation in wetlands is either of limnic in origin with water of low salt concentration, or paralic in origin with hydrological connection to the sea at the period of peat formation with higher salt consistency.

Between the two, limnic and paralic wetlands, are brackish wetlands, where seawater is occasionally pushed inland by king tides and by storms, and where diverse vegetation survives. Limnic, paralic and brackish wetlands influence characteristics of the peat and coal they produce.

Almost half of New Zealand's peatland is in the Waikato, most of which is in dairying. Peatland is also in Southland, Northland and the West Coast.

The peat layers are up to 11 metres thick and has formed over the previous 20,000 years.

Vegetation in peatland captured carbon dioxide through photosynthesis, although, as the vegetation did not fully decompose from aerobic bacteria in the water saturated soils, requiring oxygen for cellular respiration, carbon from decaying vegetation was not fully released – that would otherwise be returned to the air as carbon dioxide.

Anaerobic bacteria in wetlands, not requiring oxygen, decomposed the terrestrial matter, emitting methane gas, 'swamp gas', transforming the matter into peat, mostly decayed organic matter.

Peat is highly productive for cropping, and when land is cultivated, carbon in peat that was not released when part decomposing, combines with oxygen in the air to then be released as carbon dioxide, adding to greenhouse gas levels.

Peat may be used for domestic heating and cooking, and as boiler fuel, either in briquette or in pulverised form.

Lignite

Lignite ($C_{39}H_{35}O_{10}NS$) is a low-rank coal formed at a depth of between 200 metres and 1.5 kilometres, and at a temperature of between 25 and 45°c.

Lignite mining, Mataura, Southland

New Zealand has large reserves of lignite, a soft coal in Otago and Southland, only about 20 million years old, and is the most significant proven and recoverable energy source in New Zealand, approximately 7 billion tonnes.

This lower grade energy lignite is used for general heating, and for dairy and meat processing plants.

Coal

Coal is a hydrocarbon, although considered to be hydrogen-deficient.

New Zealand has extensive coal resources, mainly in Waikato and Taranaki regions of the North Island, and West Coast, Otago and Southland of the South Island.

In the Waikato region, main coalfields are Maramarua and Rotowaro, which produce sub-bituminous coal. The coal is used in the Glenbrook steel mill in Auckland, and for electricity transformation by the Huntly Power Station.

Taranaki coals are bituminous, although located fairly deep underground.

Coals extracted from the West Coast region, the Buller and Greymouth bituminous fields and the Inangahua sub-bituminous field, are mainly exported.

Otago and Southland regions produce sub-bituminous coal. The main coal mines are Ohai and Kaitangata.

All of these coal mines are opencast, consuming hundreds of hectares of farmland, as is depicted overleaf.

Coals may be converted into crude oil by liquification and into natural gas by gasification.

Liquefaction is soil acting as in a liquid state from an earthquake, as in the Christchurch earthquake – words are sometimes confused.

Maramarua opencast coal mine

In New Zealand, coal is not converted as crude oil and gas is in abundant supply, and the processing of coal is environmentally unsound due to the release of higher combustion emissions:

- Sulfur dioxide (SO_2)
- Nitrogen oxides (NO_x)
- Particulate matter (PM_{10})
- Carbon dioxide (CO_2)
- Mercury (Hg)
- Fly ash

Liquification of coal to crude oil for transport fuel

Coal is pulverised into a fine powder, to be able to react with hydrogen at a temperature of 450°c and at a pressure of up to 20 kPa for a period of approximately one hour, with an added solvent that assists in breaking down the coal hydrocarbon structure. The resulting liquid requires further refining for fuel products.

The hydrogen (H) bonds with the carbon (C) in the hydrogen-deficient coal to form hydrocarbons, typically octane (C_8H_{18}), which has 8 carbon atoms per molecule, at the higher range for petrol.

Conversion of coal to electric, available for transport energy

- The coal is pulverised into a fine powder.
- The powder is mixed with hot air to enable the coal to burn more efficiently.
- The hot coal is burnt in a furnace connected to a boiler for steam generation (the emission from the furnace, carbon dioxide, may be sequestered and stored deep underground in extracted petroleum reservoirs).
- The steam from the boiler spins a turbine, transforming heat energy into mechanical energy.
- The turbine powers a generator that turns mechanical energy into electric energy.
- A condenser cools the steam from the turbine, the steam condensing back into water.
- The water returns to the boiler for re-circulation.

Presently, 7% of electricity in New Zealand is transformation from sub-bituminous coal.

Chapter 19

Hydrocarbon Energy & Vehicle Combustion

Oxygen (O) is 21% of the atmosphere, nitrogen (N) 78%, carbon dioxide (CO_2) 0.04%, and nitrous oxide (N_2O) and methane (CH_4) negligible.

In combustion, fuels react with oxygen to release energy. Complete combustion in the engine requires a plentiful supply of oxygen, for all the fuel to fully react with the oxygen, emitting then as only carbon dioxide and water vapour (H_2O) from the exhaust into the air.

Both carbon dioxide and water vapour are non-toxic, although excess in the air is the effect of global warming (carbon dioxide is a greenhouse gas and water vapour a supercharger for greenhouse gases).

When incomplete combustion occurs, harmful gases are emitted from the exhaust, with also a loss of engine power.

The higher the number of carbon atoms in a hydrocarbon molecule, the higher is the ratio of carbon to hydrogen and more energy is produced.

Methane (CH_4), the main hydrocarbon in natural gas, produces the least energy as it has only one carbon atom and four hydrogen atoms, a ratio of 1-to-4, a low carbon to hydrogen ratio.

The average chemical formula for petrol is heptane (C_7H_{16}), a ratio of 7 carbon atoms to 16 hydrogen atoms, a ratio of 1-to-2.4, and for diesel, cetane ($C_{16}H_{34}$), a ratio of 1-to-2.1, producing more energy, about 15% more energy than petrol per litre.

Petrol Engines

Petrol engines generate power by injection of fuel vapour, combined with oxygen, into the combustion chamber of the cylinders.

The pistons first travel down the cylinders, drawing vapour into the cylinders as they travel. The pistons then return up, and as they travel, compress the vapour for ignition by the electric spark plugs at the top of the cylinders.

The vapour heats up as it is compressed, and in high-performance engines, before ignition by the spark plugs, spontaneous and premature ignition may occur, causing reverberations throughout the cylinders, engine 'knock' or 'pinking'.

High-performance engines are designed to compress the vapour more, which causes extra heat and thus a greater tendency to pre-ignite or detonate and 'knock', requiring premium petrol with a higher octane rating.

Such increase in power is from the fuel's higher compression ability and by engine design, not directly from the petrol.

Octane rating is combustion efficiency, the measure of how resistant petrol is to the cause of 'knocking'. Premium fuel, with a higher octane rating, is more resistant to premature detonating.

Compression ratios usually range from 8:1 to 10:1. The higher the compression ratio, 10.1, the more efficient is combustion – as the vapour is compressed, the temperature increases with a more complete combustion, producing higher torque and horsepower, and also decreasing pollutant exhaust emissions.

Compression ratio is the ratio between the largest volume of the combustion chamber with the piston at the bottom of the chamber, divided by the volume with the piston at the top.

Torque is the rotational equivalent of linear force, with a higher torque a greater initial acceleration of the vehicle. Horsepower is the work speed of moving the vehicle to its top speed, with higher revolutions per minute (RPM).

Fitting a turbocharger, the forced induction of vapour into the combustion chambers, significantly increases torque and horsepower.

The main pollutants from petrol engine exhausts are:

- Benzene (C_6H_6)
- Carbon monoxide (CO)
- Nitrogen monoxide (NO)
- Sulfur dioxide (SO_2)

and greenhouse gasses:

- Carbon dioxide (CO_2)
- Hydrocarbons (HC), mainly methane (CH_4)
- Nitrous oxide (N_2O) minimal

Diesel Engines

Diesel contains about 15% more energy than petrol as diesel has 12 to 20 carbon atoms in a molecule compared to petrol, 5 to 9 atoms. Diesel engines are 'lean-burn', as they use less fuel and more oxygen as if the same performance to petrol engines.

This enables more kilometres per litre of fuel, and as the diesel engine has higher combustion efficiency than petrol engines, the emission of the main greenhouse gas, carbon dioxide, is lower, although the concern is the emission of particulate matter.

The main pollutants from diesel engine exhausts are:

- Carbon monoxide (CO)
- Nitrogen monoxide (NO)
- Particulate matter (PM_{10}) including black carbon (BC)
- Sulfur dioxide (SO_2)

plus greenhouse gasses:

- Carbon dioxide (CO_2)
- Hydrocarbons (HC), carbon content C_{13} to C_{18}
- Nitrous oxide (N_2O) minimal

Pollution from diesel engine exhausts

Diesel, with a temperature flash point of over 52°c, is a more stable fuel in igniting and more difficult to combust than petrol, which has a flash point of minus -43°c.

Diesel engines do not have spark plugs. Air is compression heated by cylinder pistons, and then fuel vapour is injected into the heated air for ignition. Thus, the compression ratio for diesel engines is higher than for petrol engines, approximately 22:1, and operation is at a higher temperature.

Cetane rating is the quality measure of combustion for diesel fuel, a short ignition delay between injection of the fuel into the combustion chamber of the cylinders and ignition. A comparable higher cetane rating is preferred, the faster the fuel ignites for a greater power and less pollutants.

The cetane rating of fuel to be tested is comparison to the ignition delay of cetane (assigned a rating of 100) as a mixture by volume with iso-cetane (assigned a rating of 15).

Iso-cetane has the same atoms as cetane ($C_{16}H_{34}$), the mid-range of diesel fuel, although with different chemical properties.

EFSA states the cetane rating to be 51 plus and sulfur at no more than 10mg per kg. Diesel fuel does not contain the element lead (Pb), and benzene (C_6H_6) is minimal in diesel, and thus not needed to be regulated.

Diesel engines produce more torque with a greater initial acceleration than petrol engines. Petrol engines produce more horsepower with a greater top speed in each gear.

As the compression ratio of diesel engines is higher than for petrol engines, producing more torque, a stronger engine assembly is required.

Diesel engines provide most of their power from each gear on the lower range of the revolutions per minute band (RPM). The RPM for diesel-powered vehicles is normally between 1,300 and 2,000, and for petrol-powered vehicles, between 1,500 and 2,500 RPM.

Diesel vehicles are preferred in the trucking industry for the greater torque required in heavy payloads.

The opposite is for petrol engines, power builds from the increase of RPM and thus power is available in each gear for a longer period. Petrol engines produce more horsepower, and accelerates from 0 to 100 kph quicker and with higher top speed.

However, due to higher RPM, petrol engines are less fuel efficient than diesel engines.

Chapter 20

Catalytic Converter

Vehicles are now fitted with a catalytic converter, which by chemical reaction, the catalyst, mostly rare and valuable metals, convert pollutant gases, before being emitted from the exhaust system, into less harmful gases, without themselves undergoing any permanent chemical change.

Pollutant exhaust gases emitted	Conversion of pollutant gasses into
Hydrocarbons (HC) Carbon monoxide (CO) Benzene (C_6H_6) Particulate matter (PM_{10})	Oxidised with catalyst, palladium (Pd), into carbon dioxide (CO_2) and water vapour (H_2O)
Nitrogen monoxide (NO)	Broken up by 'reduction' with catalyst rhodium (Rh), and oxidised with catalyst, palladium, into nitrogen (N) and oxygen (O)
Nitrous oxide (N_2O)	Oxidised with catalyst ruthenium (Ru) into nitrogen (N) and oxygen (O)

Black carbon (BC) and other pollutant particulate matter, mainly from diesel engines, are removed by particulate filters.

Lead completely and permanently destroys catalyst activity.

Sulfur, and hydrogen sulfide gas often associated with sulfur, compete with the other exhaust pollutants for space on the catalyst surface, and thus may inhibit removal of the other pollutants.

Typical catalytic converter fitted into vehicle exhaust system.

Chapter 21

Harmful Gases Emitted from Engine Exhausts

There has been a recorded correlation between high levels of the below described harmful gases, sulfur dioxide, carbon monoxide, nitrogen dioxides, benzene and particulate matter and severe reaction from coronavirus infections. All these pollutant gases cause serious lung diseases, and coronavirus exacerbates the disease.

These gases are, however, not greenhouse gases (refer Chapter 22 for greenhouse gases).

Sulfur Dioxide (SO_2) – most sulfur dioxide produced in the atmosphere is from the combustion of fossil fuels. During vehicle combustion, sulfur (S) and associated hydrogen sulfide (H_2S) that may be present in either petrol or diesel fuel, when emitted from vehicle exhausts, oxidises to form sulfur dioxide.

Sulfur dioxide is also from the burning of tropical forests and from major volcanic eruptions.

Sulfur and hydrogen sulfide gas reduce the efficiency of catalyst converters and are corrosive to engines.

EFSA has set the sulfur limit in fuels, now down to 10 parts per million (10 mg/kg).

Sulfur dioxide is a colourless gas with a strong choking odour that smells like burnt matches. Sulfur is a health hazard, causing severe irritant to the respiratory tract, the eyes and mucous membranes, and causes severe skin burns.

Sulfur dioxide (SO_2), and nitrogen oxides (NO_X) described below, dissolve in rain droplets and form sulfuric acid (H_2SO_4), which causes smog and 'acid rain', and is very corrosive.

'Acid Rain':

- Damages respiratory tracts in humans and animals.
- Damages the cuticle of plants, the waxy layer on leaves that assists in the retaining of water.
- Leaches aluminium from soils into waterways and lakes, harmful to aquatic and insect life.
- Leaches nutrients from soils that are required for plant growth.
- Corrodes steel, etches concrete and stone, and peels paint.

Sulfur dioxide also rises into the stratosphere and combines with water vapour to form minute suspended particles, sulfuric acid aerosols.

The aerosols reflect sunlight directly back into space, creating a haze and red sunrises and sunsets, and the resultant cooling of the earth's surface.

The effect of sulfuric acid aerosols

The warming effect of greenhouse gases is around the earth, but the cooling effect of sulfuric acid aerosols is mainly in the northern hemisphere, near populated areas.

Sulfur dioxide is mostly used in the production of sulfuric acid for the chemical and pharmaceutical industries, for phosphate fertilisers and for the removal of rust. Sulfur dioxide is also used in fungicides and insecticides, and for the vulcanisation of black rubber.

Carbon Monoxide (CO) – invisible and very toxic, and is emitted from exhausts as unburnt fuel due to incomplete combustion, mainly from older vehicles, and is from both petrol and diesel engines.

Incomplete combustion occurs when the supply of oxygen to the engine is inadequate, causing carbon monoxide as a poisonous gas to be emitted from the exhaust.

The catalytic converter oxidises carbon monoxide into carbon dioxide (CO_2) and water vapour (H_2O).

Carbon monoxide is absorbed into the lungs and binds with haemoglobin in the red blood cells, reducing the capacity of the blood to carry oxygen.

Nitrogen Monoxide (NO) **and Dioxide** (NO_2), formula referring to both is nitrogen oxides (NO_x) – highly reactive and contributes to smog and to acid rain.

If excessive temperature occurs in the combustion of fuel, nitrogen (N) in the air that is with the oxygen (O) required to combust with the fuel, 'oxidises' and produces nitrogen monoxide (NO), and when emitted from the exhaust system, combines with oxygen in the air to produce nitrogen dioxide (NO_2).

Nitrogen monoxide may be emitted from both petrol and diesel engines.

Nitrogen monoxide may be reduced by modifying the combustion process, by re-circulation of the exhaust gases for further combustion, or by the catalytic converter to form nitrogen and oxygen.

A high level of nitrogen dioxide causes infections and damage to the respiratory tract, and also causes asthma. Long-term exposure causes chronic lung disease.

Benzene (C_6H_6) – is naturally in crude oil, and is emitted from petrol engine exhausts, caused by unburnt fuel from incomplete combustion.

Benzene is a hydrocarbon; is clear, colourless and has a sweet odour, and is highly flammable and volatile. The flash point is minus -11°c.

The maximum amount allowed in petrol under EFSA is 1% by volume. Only a trace of benzene is in diesel fuel.

Benzene prevents bone marrow from producing enough red blood cells, causing leukemia and anaemia, is carcinogenic (cancer forming) and damages the immune system. Benzene is in cigarette smoke.

Prior to the 1950's, benzene was used as an additive in petrol for increasing the octane rating to reduce engine 'knock'.

Benzene is used in the petrochemical industry and may be in solvents, detergents, plastics, paints, drugs, dyes, lubricants and pesticides, plus many other products.

Natural sources of benzene are from volcanoes and from forest fires.

Particulate Matter (PM_{10}) – less than 10 micro-metres in diameter, are pollutant particles suspended in air, composed of soot, dust, pollen and smoke.

Black carbon (BC), that is soot, 2.5 micro-metres in diameter, is the strongest particulate matter contributor to atmospheric warming due to the black colour absorbing incoming ultra-violet solar radiation, although as such, is not a greenhouse gas.

Driving through 'soot' on the Auckland motorway

Black carbon also influences cloud formation and rainfall.

When deposited in the cryosphere (earth's frozen region of ice and snow), the black particulate, soot, reduces albedo (the ability of white to reflect sunlight), causing melting of ice and snow and resultant rising of sea-level.

Black carbon lasts for up to a few weeks in the atmosphere before being flushed out by rain.

Incomplete combustion in vehicle diesel engines cause black carbon (and other pollutant particulates) to be emitted from exhausts, accounting for 70% of black carbon emissions in the atmosphere.

Black carbon is removed by both catalytic converters, which oxidises black carbon into carbon dioxide (CO_2) and water vapour (H_2O), and by filters.

At 2.5 micro-metres, the black carbon particles are powder like, with entry into the lungs, causing toxic build-up in the bloodstream. Also causes heart and lung disease, strokes and is carcinogenic.

Black carbon is different to carbon black. Carbon black (CB) is a manufactured product, a fine black powder used as a reinforcing agent in vehicle tyres and rubber products, and as a pigment in inks and paints.

Chapter 22

Greenhouse Gases

Causes of greenhouse gas emissions – Ministry for the Environment:

Greenhouse Gas Emissions	Percentage
Road transport	15.2
Other transport	1.5
Agriculture	50.0
Manufacturing & construction	8.5
Industrial processing	5.9
Electric generation	5.9
Waste emissions	4.1
Fugitive emissions	1.7
Other	7.2
Total	**100.0**

Fugitive gas emissions are losses of integrity, leakages during production and transport.

The atmosphere is comprised primarily of nitrogen 78%, and oxygen 21%, which are not greenhouse gases as they are transparent in the atmosphere and travel freely through the troposphere, not absorbing and then dispersing radiation heat that is outwards from the earth.

The 'greenhouse effect' is a natural process from balanced greenhouse gases rising to the troposphere, mainly carbon dioxide, 0.4% of the earth's atmosphere, to maintain a moderate temperature and to sustain living growth on earth. If it were not for this natural 'greenhouse effect', the earth's temperature would be below freezing.

However, as more greenhouse gases are emitted into the atmosphere from human activities, this causes an increase in earth's temperature, with a consequence of other climatic changes.

The greenhouse gases warming the earth's temperature originate mainly from fossil fuels and are carbon dioxide, methane and other hydrocarbons, nitrous oxide and fluorinated gases. A substantial emission of these gases has occurred since the industrial revolution, from the early 1830's.

Composition of greenhouse gases in New Zealand are:

Gas	Percentage
Carbon dioxide	46%
Methane and other hydrocarbons	42
Nitrous oxide	10
Fluorinated gases	2
Total	**100%**

Greenhouse gases are atmospheric gases in the upper troposphere, transparent to incoming solar, short-wave ultra-violet radiation to the earth's surface.

However, the long-wave infra-red that is radiated from the earth into space is mostly intercepted and absorbed by the atmospheric greenhouse gases, to then be dispersed in all directions, including back to earth, causing an increase in earth's surface temperature.

Since the 1980's, emissions of greenhouse gases are warming and thus expanding the troposphere, pushing up the tropopause, the upper limit of the troposphere. As a consequence, the stratosphere that is above the troposphere is shrinking in thickness.

Marine Transgression is occurring, the rising of sea-level, the encroaching of the foreshore onto higher land. Transgression can also be caused by earthquakes, by subsidence of land below sea-level and by tectonic plate movements.

The global sea-level began rising from the 1900's, and now averages a rise of over 3mm per year. The main cause is the emission of greenhouse gasses from fossil fuels.

The global rise in sea-level from fossil fuels results from:

- The increase in atmospheric temperature, causing the progressive melting, in both the thinning and retreating of glaciers and polar ice sheets into seawater, raising the ocean level.

- The oceans act as a buffer against an increase in atmospheric temperature. Up to 90% of additional heat on the earth's surface, caused by fossil fuels, is stored in the oceans. However, as the oceans therefore warm, the expansion of seawater occurs, raising the sea-level.

- Black carbon (BC), commonly called soot, from diesel engine exhausts, deposited in the cryosphere (earth's frozen region of ice and snow), reduces albedo (the ability of white to reflect sunlight), resulting in melting of ice and snow and the resultant rising of the sea-level.
- The decrease in land freshwater, with the transfer of water from lakes and rivers into the oceans, and the decline of vegetation able to retain soil moisture, raising the level of the sea.

Rising sea level in Thames

Carbon Dioxide (CO_2), is naturally abundant in the atmosphere, and added to this, 46% of greenhouse gas emission from human activities is carbon dioxide, mainly the combustion of fossil fuels, including from both petrol and diesel engines, overloading and putting out of kilter the composition of atmospheric gases.

Carbon dioxide is present in the atmosphere as part of the Earth's carbon cycle; oceans, soils, plants and animals, and is released into the atmosphere by natural processes, such as human and animal respiration and by volcanic eruptions.

Carbon dioxide sequestration (concealment) may reduce carbon dioxide emissions by first capturing the gas from coal-fired power plant towers before entering the atmosphere, and then storage of this gas deep underground in extracted petroleum reservoirs.

As the number of carbon atoms increase in heavier fossil fuels, the ratio of carbon (C) to hydrogen atoms (H) increase. When carbon atoms burn they oxidise into carbon dioxide (CO_2), and hydrogen atoms oxidise into water vapour (H_2O), the reason why heavier fuels, like coal, tend to have higher emissions of carbon dioxide per unit of energy.

The table below shows for fuel types, the number of carbon atoms per molecule and their kilograms (Kg) of carbon dioxide (CO_2) emitted per gigajoule (Gj) of energy produced.

Fuel Type	Carbon Atoms per Molecule	Kg CO_2 per Gj of Energy
Natural gas	1-2	240
LPG	2-4	290
Petrol	5-9	325
Diesel	12-20	340
Coal	> 200	435

Excess exposure to carbon dioxide is a health hazard, with symptoms: headaches, dizziness, increased heart rate, high blood pressure, asphyxia and convulsions, caused by, for example, running an engine in an enclosed garage.

Methane (CH_4) is the principle hydrocarbon in natural gas, is colourless, odourless and flammable, and occurs in nature, 1.7 parts per million.

Natural emissions are in wetlands, the anaerobic bacteria decomposition of vegetation and animal matter that is submerged in water and thus oxygen deprived, producing methane, 'swamp gas'.

Methane and other hydrocarbons account for 42% of greenhouse gas emissions from human activities, such as:

- livestock farming — animal digestive processes; over 80% of the methane produced is from this source,
- leakage during production and transportation of gas,
- the incomplete combustion from vehicle engines, and from the burning of forests and in wood fuel for domestic fires,
- from domestic and industrial waste-water treatment plants, and
- the anaerobic bacteria decomposition of organic landfill waste.

Landfill sites tend to be anaerobic, oxygen deprived, emitting methane gas. By injecting air and moisture into the landfill, anaerobic bacteria are destroyed within a few weeks and replaced by aerobic bacteria that release less harmful carbon dioxide, although still a greenhouse gas.

Methane is approximately 25 times more potent than carbon dioxide in absorbing and then scattering long-wave infra-red radiation from earth, although remains in the atmosphere for less — 12 years or so compared to carbon dioxide for over 100 years.

High levels of methane result in mood changes, slurred speech, vision problems, memory loss, nausea and headaches.

Nitrous Oxide (N_2O), 'laughing gas', is a colourless and non-flammable gas with a slight metallic scent, which also lingers in the atmosphere for over 100 years, and is more potent than carbon dioxide in warming the atmosphere.

Nitrous oxide is produced from natural sources, mainly from aerobic bacteria in the breaking down of nitrogen in soils.

About 40% of nitrous oxide in the atmosphere is from human activities, (and accounts for 10% of total human greenhouse gas emissions), which is mainly from applying synthetic nitrogen fertilisers to the land for the growth of crops and pastures.

The emission of nitrous oxide is also from incomplete combustion of fossil fuels, mainly from coal but also from vehicle fuels, both petrol and diesel.

The catalytic converter in vehicle exhaust systems, which by catalyst ruthenium (Ru), oxidises nitrous oxide (N_2O) into elements nitrogen (N) and oxygen (O). Nitrous oxide may also be decomposed at a temperature of around 600°c into nitrogen and oxygen.

Racing cars often use nitrous oxide (N_2O) as fuel, which is one oxygen atom and two nitrogen, whereas petrol (C_5H_{12}) to (C_9H_{20}) is composed of just 0.1 to 1.0% of oxygen.

This increased level of oxygen in nitrous oxide enhances the speed of oxidisation, with the resultant greater injection of fuel into the combustion chamber, producing an increase of power.

Nitrous oxide chemically reacts with ozone in the stratosphere to form nitrogen monoxide (NO), a dangerous molecule that destroys the ozone layer (Chapter 23).

Fluorinated Gases account for 2% of human greenhouse gas emissions. There are no natural sources of these gases, only from human-related activities.

One use of fluorinated gases is as refrigerants in vehicle and building air-conditioning systems, and it can be emitted as gas leak pollutants.

Hydro-fluoro-carbons (HFC) – hydrogen (H), fluorine (F) and carbon (C) atoms, were developed as refrigerants for the replacement of hydro-chloro-fluoro-carbons (HCFC) – hydrogen, chlorine, fluorine and carbon atoms, as when chlorine (Cl) entered the stratosphere, it reacted with ozone and destroyed the earth's protective 'ozone layer'.

However, HFC contains carbon atoms, which when released into the air, combine with oxygen atoms to form carbon dioxide, creating high global warming potential.

Newly developed refrigerants hydro-fluoro-olefins (HFO) – hydrogen (H), fluorine (F) and olefins (C_nH_{2n}) replaces carbon atoms to have less impact on global warming.

Thus, the effect of fluorinated gases on both global warming and depletion of the 'ozone layer' will gradually lessen as the refrigerants, HFO, are progressively used.

Chapter 23

Ozone Layers

There are two 'ozone layers'.

- Ozone (O_3) is a molecule of three oxygen atoms, and collective ozone molecules form the higher altitude, protective stratosphere 'ozone layer'.
- The lower, ground level troposphere 'ozone layer' is formed by human-created pollutants.

Stratosphere 'Ozone Layer'

The Stratosphere 'Ozone Layer' that is above the troposphere encasing the earth, shields the planet from harmful ultra-violet-B and C radiation rays generated from the sun. The upper stratopause boundary extends to about 50 kilometres from earth.

Depletion of the stratosphere 'ozone layer' from pollutants occurs globally, although severe depletion is from the mid-latitudes to the Antarctic, and the effect on ozone depletion is upwards of 100 years.

New Zealand is one of the countries affected by depletion of the 'ozone layer', which occurs annually from late winter to December. Called an 'ozone hole', is technically an ozone depletion of yearly intensity of up to 70%.

The cause of ozone depletion is from nitrous oxide, and from chlorine and bromine:

Nitrous oxide

Nitrous oxide (N_2O) rises to the stratosphere and chemically reacts with ozone, oxidises, to form nitrogen monoxide (NO), a dangerous molecule that destroys the 'ozone layer'.

About 40% of nitrous oxide in the atmosphere is from human activities: including the emission from incomplete combustion of vehicle fuels, both petrol and diesel engines.

Nitrous oxide is also a greenhouse gas.

Chlorine and bromine

Chlorine in HCFC, a refrigerant in older vehicle air-conditioning systems, and previously a propellant in aerosol spray cans, also rises to the stratosphere and reacts with ozone to destroy the earth's protective 'ozone layer'.

Both chlorine (Cl) and bromine (Br) are used as disinfectants and bleaching agents.

As stated in Chapter 22, since the 1980's, emissions of greenhouse gases are warming and thus expanding the troposphere, pushing up the tropopause boundary, the upper limit of the troposphere. As a consequence, the stratosphere, the ozone layer that is above the troposphere, is shrinking in depth exacerbating the effect from depletion of the ozone protective layer.

Depleting of the 'ozone layer' lowers the yield of agricultural crops by limiting photosynthesis, the process of vegetation using sunlight to take in carbon dioxide and water to synthesize nutrients. Which in turn limits the ability of vegetation to act as 'carbon sinks', the take-up of carbon dioxide that is produced both naturally and by human activities, influencing the warming of the earth's surface.

Thus, depleting the 'ozone layer' indirectly raises the earth's temperature by increasing atmospheric carbon dioxide, the main cause of the greenhouse effect.

Ultra-violet radiation striking the earth due to the thinning of the 'ozone layer' causes an increase incidence of skin cancer and cataracts, and affects the human immune system.

Troposphere 'Ozone Layer'

The troposphere 'ozone layer' is closer to the earth's surface, and is a pollutant layer causing city smog.

The upper boundary of the troposphere above the earth's surface, the tropopause, varies from under 10 kilometres at the north and south poles to 20 kilometres at the equator.

In the troposphere, temperatures drop with increasing altitude and in the higher stratosphere temperatures rise with increasing altitude. Because of this gradual temperature stratification, there is no turbulence between the lower troposphere and the higher stratosphere.

Smog is categorised as either summer smog or winter smog.

Summer Smog

An 'ozone layer' at ground level during summer months, smog, is a pollutant gas created by photochemical reaction, the absorption of energy from sunlight by vehicle exhaust emissions and industrial and chemical plant emissions, mainly nitrogen dioxide (NO_2) and particulate matter (PM_{10}) from vehicle exhaust emissions, and combining these emissions with volatile organic compounds (VOC).

VOC may be in and released by many products:

VOC Compounds From:	Used In:
Acetone	Nail polish remover
Acetic acid	Vinegar
Alcohol	Solvents
Butanal	Candles, barbecues and gas stoves
Carbon disulfide	May be in chlorinated tap water
Ethanol	Cleaning products
Formaldehyde	Plastics and lacquers
Methylene chloride	Paint removers and aerosol solvents

Summer ozone smog is especially worse on hot sunny days in the city, due to the photochemical reaction of energy from the sun. The ozone level, with its resultant smog, reaches its peak during mid to late afternoon, from the build-up of exhaust fumes and heat of the sun, and lasts for up to 4 hours in the atmosphere.

High concentrations of ozone near ground level is harmful to both people and vegetation. Ozone smog irritates the eyes, nose and throat, and inflames the respiratory system and aggravates asthma.

Results from about the 45 air monitoring sites around Auckland verifies there is photochemical reaction occurring with fossil fuel emissions and VOC, and that there is an elevated level of this pollutant ozone during summer months.

Winter Smog

During winter, two incidences combine to cause ozone winter smog:

1. **An increase in fossil fuel usage**
   During the winter, the cooler temperature causes an increase in fossil fuel and wood combustion to heat homes and buildings (except where this is banned, such as in parts of Rotorua due to smog settling in the formed volcanic crater in which the city lies). These increased combustion pollutant emissions are released into the atmosphere. Plus the combination of pollutant emissions from vehicles.

2. **Natural phenomenon of temperature inversion**
   Temperature inversion is where, on a windless night, air that contains combustion pollutants and VOC becomes cooler by conduction with the earth's surface, which has cooled off rapidly by emitting long-wave infra-red radiation into space, and this cooler air with pollutants and VOC is trapped by the less dense and lighter warm air above, like a lid.

As a consequence, this pollutant mixture cannot be released into the higher troposphere, which causes 'ozone smog' near earth's surface.

Christchurch winter smog

Hamilton's Fog

The fog in Hamilton is 'radiation fog', not a pollutant 'ozone layer' smog. The fog forms on windless nights when heat quickly radiates into space. As the air cools closer to the earth, it causes moisture in the air to condense, forming fog.

If there is no wind during the day and with a low angle of the winter sun, the fog can persist until the afternoon.

Hamilton is prone to radiation fog, as Waikato is a large flat area where heat radiates away quickly, it is inland from temperature moderation of the sea, and the Waikato River provides a source of moisture.

Chapter 24

Regulations

The Government organisation, New Zealand Petroleum & Minerals (NZP&M), is a branch of the Ministry of Business, Innovation and Employment (MBIE) that administers the Crown Minerals Act 1991 and related regulations (and its predecessor, the Petroleum Act 1937) on behalf of Minister of Energy and Resources.

NZP&M regulates the issuing of prospecting and mining permits to New Zealand and foreign petroleum companies of Crown-owned petroleum resources. The Crown receives royalties from the companies.

Criteria for granting of exploration permits is based on the petroleum companies' technical and financial capabilities to meet required health, safety and environmental requirements.

The Crown Minerals (Petroleum) Amendment Act 2018, states, effective from 12th November 2018, new applications will only be approved for petroleum prospecting and mining permits covering the on-shore Taranaki region. Existing permits outside this area are unaffected and may continue mining.

Thus, all new off-shore mining permits are now prohibited, as well as those on-shore outside the Taranaki Region.

The petroleum companies also need a marine consent, which is issued by the Environmental Protection Authority (EPA), the Government agency responsible for regulating the effects of restricted activities such as fossil fuel drilling.

Maritime New Zealand (MNZ) is responsible for ensuring operators have plans in place to respond to a potential oil blow-out or spill.

The Crown owns all petroleum resources in New Zealand as well as maritime jurisdiction, administered under the authority of the United Nations Convention of the Law of the Sea (UNCOLS), for the exploration and production of oil and gas, extending to the Treaty Line of the continental shelf, Zealandia, on which New Zealand lies.

The 4 zones, with their rights, stretching from the shores of New Zealand are:

- **Territorial Waters:** New Zealand has full sovereign rights in the Territorial Waters from under the seabed to the airspace above, extending 12 nautical miles out from the low tide mark of New Zealand shores, including that of the islands, Campbell Island, Auckland Island, Antipodes Island, Bounty Island, Chatham Island and Raoul Island — one nautical mile is equivalent to 1.85 kilometres.

- **Contiguous Zone:** New Zealand has sovereign rights to prevent violations of its environmental, customs, fiscal and immigration laws, and the control of all economic resources in the Contiguous Zone (including petroleum reservoirs) extending from the Territorial Waters to 24 nautical miles from land, including from those islands.

- **Exclusive Economic Zone:** New Zealand has sovereign rights to the control of all economic resources in the Exclusive Economic Zone (including petroleum reservoirs) extending from the Contiguous Zone to 200 nautical miles from land, including from those islands.

- **Continental Shelf:** New Zealand has sovereign rights to resources on and under the seabed to the Continental Shelf, Zealandia (including petroleum reservoirs), from the Exclusive Economic Zone out to the Continental Shelf Treaty Line with Australia.

The Exclusive Economic Zone and Continental Shelf (Environmental Effects) Act 2012 promotes the sustainable management of natural resources and the protection of the environment from pollution.

However, the Government has now legislated against future mining for fossil fuels in all four zones for both local use and export of fossil fuels.

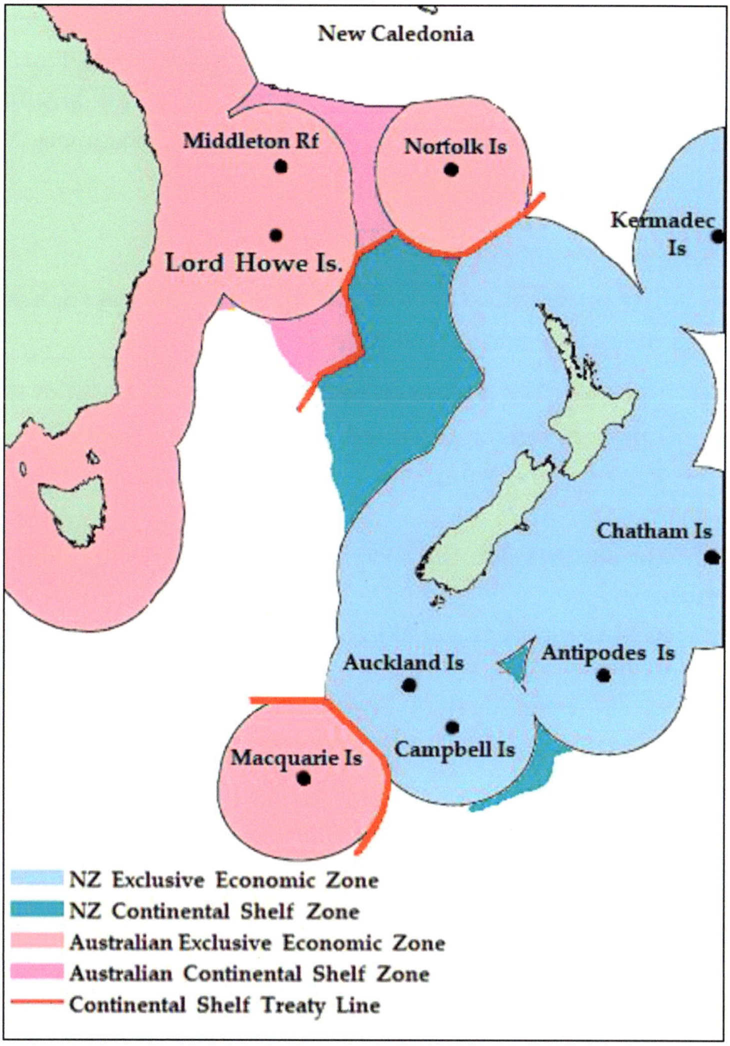

New Caledonia, at the top of the map, is a territory of France with its own Exclusive Economic Zone around those islands.

Fuel quality is governed by the Engine Fuel Specifications Regulations 2011 and Amendments, and is administered by MBIE routinely testing petrol and diesel samples from around the country, to ensure that the fuels available to consumers comply with those regulations.

The Regulations set out:

- The minimum standards of fuel performance, such as the octane and cetane ratings.
- Limiting fuel composition that could be harmful to the environment and to public health, such as lead, benzene and sulfur.

Other Government acts and their regulations that come into play are:

Resource Management Act 1991

If fossil fuel drilling activity is in Territorial Waters, which extends from the sea-shore out to the 12 nautical mile limit, Regional Councils are responsible for managing the environmental impact. Operators are required to obtain a resource consent.

Maritime Transport Act 1994

To regulate maritime activities and the marine environment in New Zealand waters to the Exclusive Economic Zone, and to the Continental Shelf Zone as permitted under international law.

Gas Act 1992

To provide for the regulation, supply and use of gas in New Zealand.

Hazardous Substances and New Organisms Act 1996

To protect the environment, and the health and safety of citizens and communities, by preventing and managing the adverse effects of hazardous substances and new organisms.

Marine and Coastal Area Act 2011

States the land from high water mark to 12 nautical miles out to sea is incapable of ownership, including that of ownership to the Government.

Health and Safety at Work Act 2015

The Regulation to the Act, the Health and Safety at Work (Petroleum Exploration and Extraction) Regulations 2016 is for the health and safety in exploration and extraction of fossil fuels.

Climate Change Response (Zero Carbon) Amendment Act 2019

To implement climate change policies in accordance with the global effort under the Paris Agreement, to limit the global average temperature increase to 1.5°c above pre-industrial levels, and to set a domestic greenhouse gas emission reduction target to zero by the year 2050.

Chapter 25

Fuel Taxes

Taxes and levies at source (when fuel is purchased by consumers) and Road User Charges (RUC) are the primary means of funding for building and maintaining road networks and for related transport facilities. The government receives in revenue approximately 4 billion dollars per year in fuel charges, or nearly $1,200 per vehicle licence holder in New Zealand.

Motor vehicles are a way of life, a necessity, to the extent that with an increase in fuel price, this reflects in a minimal reduction in national fuel consumption. As petrol or diesel is a necessity, the change in the demand for fuel after an increase or decrease in price is minimal. If the price of fuel increases due to an increase in taxes, production cost rise or from an interruption of supply, the effect on demand for fuel is fairly minimal, or is inelastic.

Luxury items or unnecessary items are elastic, in that a variance in price causes a significance and opposite variance in demand.

All fuel taxes are regressive taxes in that, as fuel is an essential expenditure, for the same quantity of fuel consumed, the poorer households spend a greater percentage of their income on fuel than those of richer households. Thus, the burden of financing the transport infrastructure by fuel charges bear more heavily on the bottom quintile of households (that is, the lower 20% of households as regards income), calculated as between 10 and 20% more than by the top quintile of households.

Adding to this of taxing for roading relatively more to the lower quintile, this group generally has less efficient vehicles, causing a higher tax take per kilometre travelled for the lower quintile of households.

Taxes, duties and levies on fuel in cents per litre:

	Petrol	Diesel
Fuel Excise Duty	70.0	0
Local Authorities Fuel Tax	0.7	0.3
Auckland Regional Fuel Tax	13.3	13.3
Petrol Monitoring Levy	0.6	0.6
ACC Levy	6.0	0
Emissions Trading Levy	20.0	23.0
Total, excluding GST	**$1.10.6**	37.2

- Auckland Regional Fuel Tax is included in the table. Deduct 13.3 cents per litre for the rest of the country.

- Accident Compensation Corporation Levy (ACC) on diesel vehicles is paid when registering the vehicle.

- Due to the war in Ukraine and the consequent increase in petrol prices, for 3 months from 15th March 2022, Fuel Excise Duty (FED) has been reduced by 25c per litre from 70c to 45c. And now the 25c reduction has been extended to January 2023.

- Diesel is not subject to FED as taxes on diesel vehicles are subject to RUC.

- New Zealand operates an Emissions Trading System (ETS) under the Climate Change Response Act 2002, a levy on greenhouse gas emissions, carbon dioxide, methane and nitrous oxide, known as a 'carbon tax'.

  The Scheme was imposed for financing of research and in development of renewable energy, and also to encourage environmentally sustainable behaviour of firms and the public.

  Although, as fuel taxes are inelastic in nature, this levy is not effective in its purpose for environmental behaviour, and only the top quintile of households able to take-up the subsidy for purchase of EV vehicles.

  'Carbon credits' are settled, one 'carbon credit' equivalent to one tonne of greenhouse gases, specified as one NZU (NZ Unit). The 'carbon credits' are tradable, allowing the purchase holder of the 'credit', the right to emit one tonne of carbon dioxide or the equivalent weight of another greenhouse gas.

The price of one NZU traded at $85 on 31st October 2022. One year previous, the trade price was $65 per NZU.

The $85 price of one NZU extrapolates to a levy of about 20c per litre for petrol at the pump and 23c for diesel.

By the year 2026, the price of one NZU is expected to rise to $110, with a levy on petrol of over 25c per litre and diesel of over 29c per litre.

The Emission Trading Levy on fuel used for international aviation and marine transport is exempt.

- In addition, all fuels are subject to 15% goods and services tax (GST), including that on taxes and levies.

- If the retail price of petrol is $3.00 per litre, GST would be 39 cents, and added fuel taxes, levies and duties of $1.11, would be a total Government take of approximately $1.50 per litre for the Auckland Region, 50% of price at the pump.

- Plus, in additional to the above taxes, levies and duties, the Government receives royalties from petroleum exploration and production companies, and taxes from nationwide petroleum distribution companies and from individual service stations proprietors.

Split of fuel prices in New Zealand is approximately:

Percentage	Type of Costs
40%	Singapore spot purchase price, including freight and insurance
50%	New Zealand Government taxes, levies and duties
10%	Importer's distribution and service station retailing costs and with their profit margins

Regional Fuel Tax

13.3 cents per litre (plus GST). RFT was imposed within the Auckland City Council region (excluding Great Barrier Island) for a period of 10 years from 1st July 2018, to fund transport projects that would otherwise be delayed or not funded. The revenue generated from RFT is approximately 1.5 million dollars per year.

RFT is intended to be applied only where fuel is for on-road vehicles. When fuel has been used for off-road activities, a rebate may be claimed, although a small amount of this revenue is lost through irregular accounting.

The revenue generated from RFT is towards:

- Improvement in public transport – bus service lanes, ferry services, electric trains and more Park and Rides.
- Extension of walking and cycling tracks.
- Improvement in road safety.
- Improvement of arterial roads.
- Roading network for residential and commercial growth regions.

Bus lane and cycling track funded by RFT

Suburban roading infrastructure is mostly funded by local and regional councils and by property developers, and regional roads, highways and motorways by the Government.

Toll Charges

Toll charges, a tax used for specific roading infrastructure, is monitored by NZ Transport Agency.

There are 3 toll roads in New Zealand

Road User Charges

RUC were introduced in 1977 to replace the conventional diesel fuel tax at source that had become difficult to administer due to between 30 and 40% of diesel fuel being used by off-road vehicles (tractors, forklifts and bulldozers), in ships and for industrial and commercial activities. This off-road use was not associated with roading network development and maintenance, and heavy vehicles powered by diesel needed to pay for their fair share of road damage.

Off-road and other users of diesel previously had to pay the diesel tax at source and then request a refund for off-road use.

Light vehicles, under 3.5 tonnes and powered by petrol or LPG, contribute to the FED and other taxes and levies as detailed above when re-fuelling. Light vehicles that are powered by diesel or bio-diesel now pay RUC by purchasing a Distance Licence to be displayed on the vehicle's windscreen.

Distance Licence

Light vehicles powered with petrol or with diesel pay about the same amount in FED, taxes and levies or in RUC for the same quantity of fuel purchased. However, diesel engines are more fuel efficient than petrol engines, and thus per kilometre pay less for roading network development and maintenance.

All heavy vehicles (over 3.5 tonnes), including those that run on petrol, are subject to RUC as they all cause damage to the roads, the heavier the vehicle the more damage and with higher RUC.

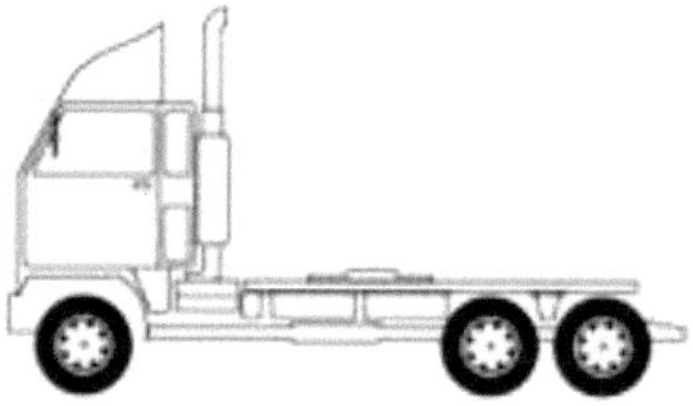

Trucks 3.5 to 12 tonnes — $110 RUC per 1,000 km

Trucks 12 to 18 tonnes — $356 RUC per 1,000 km

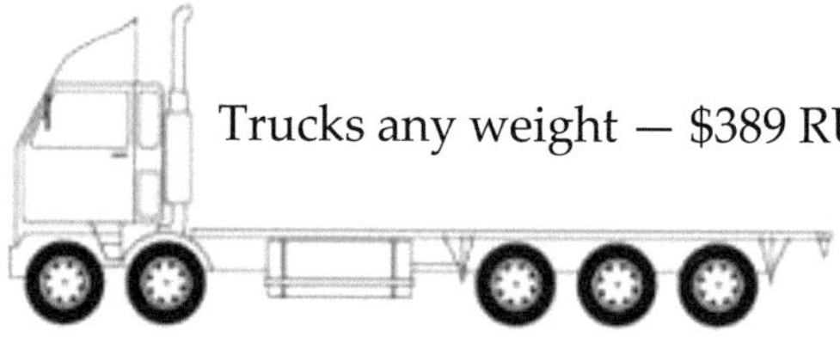

Trucks any weight — $389 RUC per 1,000 km

All vehicles that are required to pay RUC must be fitted with an approved electronic distance recorder for the distance travelled.

Light electric vehicles are exempt from RUC until 31st March 2024 and heavy electric vehicles until 31st December 2025.

Chapter 26

Import & Export of Oil

The high quality 'sweet' and 'light' oil that predominates in New Zealand, a relatively pure oil, obtains premium prices on the international market, and approximately 93% has previously been exported.

Petroleum was within New Zealand's top ten of exported products.

New Zealand has adequate crude oil reservoirs to be self-sufficient in domestic demand and to be able to export, although has recently chosen the path of prohibiting all new off-shore exploration and production of petroleum and restricting on-shore to just the Taranaki region.

Ocean oil tanker

Year 2021 – before Marsden Point Refinery was de-commissioned.

The importing of crude oil, nearly 27 million barrels, to be refined at Marsden Point Refinery:

Country	%	MB
United Arab Emirates	79	21.0
Malaysia	9	2.5
Saudi Arabia	9	2.5
Russia	3	1.0

The importing of refined petroleum products, approximately 21 million barrels:

Country	%	MB
South Korea	43	9.0
Singapore	40	8.5
Other countries	17	3.5

During 2021, 6.4 million barrels of high quality crude oil were exported, manly to Australia, and under 10% to Singapore and to United States.

Marsden Point Refinery was set-up to process heavier crude oil than that produced in New Zealand; hence the export of high quality New Zealand crude oil for top dollar and the import of heavier crude to be processed.

Chapter 27

Marsden Point Refinery

New Zealand's only crude oil refinery, Marsden Point Refinery at Whangarei, opened in May 1964 with a capacity of up to 24 million barrels per year (Pj 145), and was until recently operated by Refining NZ.

Marsden Point was chosen, as the site has a deep-water harbour close to the main market of Auckland, and is in a low earthquake risk zone, Northland. At inception, Marsden Point Refinery was considered to be one of the most versatile and sophisticated refineries in the world.

Over the intervening years, the refinery expanded and upgraded to meet quantity and environmental demands.

Production capacity more than doubled to 50 million barrels per year (Pj 300) since refinery inception, and supplied up to 60% of New Zealand's refined fuel; petrol and diesel, plus aviation and maritime fuel, processed mainly from crude oil imports.

A 170 kilometre pipeline from Marsden Point Refinery to the South Auckland terminal at Wiri supplies petrol, diesel and aviation fuel to Auckland and the Waikato, pumped at a flow rate of about 320,000 litres per hour.

Northland distribution is by road tankers, and to the country south of Waikato by two coastal tankers and then by road tankers.

Road tankers distribute close on 6 billion litres of petroleum per year to approximately 1,200 service stations throughout New Zealand.

Gull NZ imports refined fuel to its storage tanks at Tauranga, and then by road tankers to their retail outlets around the North Island.

Upgrading of the Marsden Point Refinery	
1985	Hydro-cracker unit installed for heavier crude oil processing, environmental treatment units installed, and the pipeline to Auckland laid.
2005	De-sulfuriser for diesel, and benzene remover for petrol installed.
2009	Increased crude oil capacity by installing distillation unit.
2015	Installation of a continuous catalytic reactor for increased production and environmental performance.
2018	Full refurbishment of the refinery's hydrogen manufacturing unit.

Diminishing operating margins at Marsden Point Refinery were caused by mega refineries with their economies of scale, operating in China, South Korea, Singapore and India, and of late by Covid-19 reducing demand of vehicle, aviation and maritime fuel.

Marsden Point Refinery

The refinery operations are being de-commissioned, and the assets of Refining NZ utilised by Channel Infrastructure for the refinery to become an import and distribution terminal of refined transport fuel, the last month of operation, March 2022, although complete terminal upgrade and de-commissioning will take a number of years.

The supply of all transport fuels, the lifeblood of any country, is now to be relied upon in New Zealand by the world economy, affected by embargoes and wars (as what is happening at the time of writing).

Petroleum products will to be sourced from multiple countries to minimise supply risks. Historically, prior to Marsden Point Refinery inception, all fuels were imported into New Zealand, independently by oil companies.

Chapter 28

International Trade

The International Energy Agency (IEA), a subsidiary of the Organisation for Economic Co-operation and Development (OECD), ensures future petroleum price stability for countries in response to the oil crisis in 1973.

The 1973 oil crisis occurred when the Organization of Petroleum Exporting Countries members led by Saudi Arabia proclaimed an oil embargo against those countries supporting Israel in the Arab-Israel war. The embargo caused the oil price to increase by nearly 300%.

To enable stability, the 31 member countries of the IEA (number of countries continually vary) are required to maintain 'oil stocks' equivalent to at least 90 days of oil consumption, either in their own country or from 'Stock Tickets' obtained from member countries.

In New Zealand, reserves fluctuate of over 100 days oil supply, varying on oil tankers off-loading, over the 90 requirement.

New Zealand has agreements with the United Kingdom, Japan, Denmark, Spain and the Netherlands to enable 'oil stocks' held in those countries, or 'Stock Tickets', to be counted towards New Zealand's IEA obligations.

In New Zealand, crude oil prices are generally referenced to the spot market price in Singapore as this country is one of the world's largest refining centres and is the country where much of New Zealand's petroleum has been traded.

Other countries may reference their crude oil prices to:

- The cartel, the Organization of Petroleum Exporting Countries (OPEC) — the price is the average from oil fields in 15 countries (countries vary), Algeria, Angola, Ecuador, Equatorial Guinea, Gabon, Iran, Iraq, Kuwait, Libya, Nigeria, Qatar, Republic of the Congo, Saudi Arabia, United Arab Emirates and Venezuela.

  Venezuela has the world's largest known oil reserves, and OPEC controls over 50% of world's oil supply and has 80% of proven oil reserves.

  The cartel aims to exert control in oil supply, to be able to set the price of oil on the world market.

- Brent Crude, a 'sweet' and 'light' crude oil from fields between Scotland and Norway in the North Sea, sent to refineries in Britain, supplies petroleum products to most of Europe.

- West Texas Intermediate (WTI), prices from oil fields mostly in Texas, supplying oil and petrol to North America. A 'sweet' and 'light' crude oil of high quality.

- Dubai Crude, prices from oil fields in Dubai, supplying crude oil to Asian refineries. A 'light' and 'medium sour' oil.

The other big player is Russia, which has the world's largest gas reserve, the 2nd largest coal reserve and the 8th largest oil reserve. Over 40% of Europe's fuel is imported from Russia.

Crude oil trading

Purchase of oil may be spot contracts or future contracts.

Spot contracts is the current market price of oil, and future contracts the price buyers are willing to pay for oil on a set delivery date at a set price in the future. The futures price is no guarantee the market price of oil will actually be that price at the delivery date. It is the future anticipated price by the purchaser on which they pay.

Most commodity contracts are traded on the spot market, which takes effect immediately. Money is exchanged, and the purchaser accepts delivery of the goods. Oil is usually traded in future contracts due to the logistics of transporting the oil.

A contract is signed with the investor, the dealer, for the end-user to purchase a certain quantity of oil at a predetermined price on a predetermined delivery date.

The end-users of oil purchases on the futures market lock-in a purchase price and the dealer takes a gamble to make a profit.

The price difference between the spot market and the futures market indicates the anticipation if, in the foreseeable future, the price of oil will increase or decrease. If futures prices are higher than spot prices, purchasers anticipate the market price will rise.

When the futures price is above the spot price, this is termed contango, and when futures price is below, is termed backwardation.

Crude oil moves through volatility of supply and demand, affected by worldwide output as well as global economic prosperity, embargoes, price-fixing and cold-war disruptions, now what New Zealand will be more part of, due to the Labour Government accepting decommissioning Marsden Point Refinery, effective March 2022.

The price of crude oil rose during 2007–08 to over $140 US per barrel or about 90 cents per litre, caused by a strong demand and a stagnating world production.

Then, a few years later, 2014–16, a supply glut of crude oil from a failure of an expected global growth caused the price per barrel to collapse to below $40 US or 25 cents per litre of crude oil.

The Ukraine-Russian armed conflict has pushed OPEC price per barrel to $126 and Brent to $123 in early March 2022, (although since lowered) the conflict because of Ukraine wanting to join NATO and thus allowing NATO and American military forces (as is their intent) to be on the border with Russia, also the conflict due to the wanting of annexation by Russia of the more pro-Russian, independent states, Donetsk and Luhansk.

Elements and compounds associated with crude oil and gas

Benzene (C_6H_6)
Barium sulfate ($BaSO_4$)
Black carbon (BC)
Bromine (Br)
Calcium chloride ($CaCl_2$)
Carbon (C)
Carbon black (CB)
Carbon dioxide (CO_2)
Carbon monoxide (CO)
Chlorine (Cl)
Ethanol (C_2H_5OH)
Fluorine (F)
Gas hydrate ($C_4H_{18}O$)
Hydrogen (H)
Hydrocarbons (HC)
Hydro-chloro-fluoro-carbons (HCFC)
Hydro-fluoro-carbons (HFC)
Hydro-fluoro-olefins (HFO)
Hydrogen gas (H_2)
Hydrogen sulfide (H_2S)
Lead (Pb)
Lignite ($C_{39}H_{35}O_{10}NS$)
Magnesium chloride ($MgCl_2$)
Methane (CH_4)
Methanol (CH_3OH)
Mercury (Hg)

Natural gas (CH_4)
Nitrogen (N)
Nitrogen dioxide (NO_2)
Nitrogen monoxide (NO)
Nitrogen oxides (NO_x)
Nitrous oxide (N_2O)
Olefins (C_nH_{2n})
Oxygen (O)
Ozone (O_3)
Particulate matter (PM_{10})
Palladium (Pd)
Platinum (Pt)
Ruthenium (Ru)
Sodium chloride (NaCl)
Sodium plumbite (Na_2O_3Pb)
Soot ($PM_{2.5}$)
Sulfur (S)
Sulfuric acid **(**H_2SO_4)
Sulfur dioxide (SO_2)
Volatile organic compounds (VOC)
Water vapour (H_2O)

Abbreviations in the Book

API	American Petroleum Institute
B7	Bio-diesel of 7%
CNG	Compressed natural gas
E10	Bio-ethanol of 10%
EFSA	Engine Fuel Specifications Amendment
EPA	Environmental Protection Authority

ETS	Emissions trading system
EV	Electric vehicles
FCEV	Fuel cell electric vehicles
FED	Fuel Excise Duty
IEA	International Energy Agency
kPa	Kilopascals
LNG	Liquefied natural gas
LPG	Liquid petroleum gas
MB	Thousands of crude oil barrels
MBBl	Thousands of crude oil barrels
MBIE	Ministry of Business, Innovation & Employment
MMBBl	Millions of crude oil barrels
MNZ	Maritime New Zealand
MPa	Megapascals
NZP&M	New Zealand Petroleum & Minerals
NZU	New Zealand Unit (one tonne of 'carbon credit')
OECD	Organisation for Economic Co-operation & Development
OPEC	Organization of Petroleum Exporting Countries
Pj	Petajoules
RPM	Revolutions per minute
RUC	Road User Charges
SAF	Sustainable Aviation Fuel
TB	Thousands of crude oil barrels
UNCOLS	United Nations Convention of the Law of the Sea
WTI	West Texas Intermediate

Made in the USA
Las Vegas, NV
04 November 2026

57425106R10083